Billion Dollar Napkin

PROVEN PATHWAYS TO COMMERCIALISE
YOUR INTELLECTUAL PROPERTY

Daniel J O'Connor

B.Bus, MBA, CPM, FAICD, AIMM, MAIM, MAIeX AIMCM.
Consultant Principal – **incub8IP.com**

First edition 2016

National Library of Australia

Cataloguing–in–publication entry:

O'Connor, Daniel J. 1958-

Billion Dollar Napkin: Proven Pathways to Commercialise your Intellectual Property/ Daniel J. O'Connor.

1ˢᵗ Ed.

ISBN 978-0-646-93791-5

       Intellectual Property

       Inventions and Innovation

       Commercialisation, Commercialization

       Business Development

Cover Design and typeset by Daniel O'Connor

Attadale, Perth Western Australia 6156

For reordering information:

Phone: +61 417 956 433

www.billion-dollar-napkin.com

CONTENTS

Chapter 1 – Myths Models & Rules 1

Understanding the Rules 2

    Get Realistic..2

    Emotional Commitment 3

Build the Business Model First 3

    Project Focus..6

    Rule of Costs..7

    Offer Commercial Products Only 7

Stepping Away..8

    Packaging Your Innovation......................................9

Spin-Out..9

Spin-In...10

    Presenting what investors want to buy...................10

Prove the Concept..11

    Other Factors 12

    Targeting the Investment Profile.............................12

    Reframe Your Pitch 14

What Most Investors Will Want...................................15

    How to ask for more and get it...............................16

Stages and Phases 18

    The Idea...18

    The Concept..20

    The Project...21

    The Company..24

    The Exit...25

The Phases 26

Chapter 2 – The Six Roadblocks **31**

Structure & Ownership 32

Proof of Concept 36

Funding 38

Focus 43

Distribution & Scale-up 44

Exit Options 46

Chapter 3 – The Seven Deal-Killers **47**

Rushing to Patent 48

No Compelling Pitch 52

Wrong Funding and Structure model 54

Too Many Concurrent Projects 55

Undefined Areas of Expertise 56

Complexities & Distractions 57

Not Enough Funding 58

Becoming the hunter 62

Chapter 4 – The Funding Search **67**

What do you need to pitch? 73

    Funding Increments 82

    One Verses Many 82

    Timing of the Draw-Down 83

Financial Offsets 84

The Cornerstone Investor Model 84

Understanding other Transaction Currencies 85

    Acquiring IP without cash 87

    Situations 87

Types of Currencies 88

Understanding Risk 90

How I make investors 93

Chapter 5 – Finance **97**

Allocation of Capital 98

Raising the Capital 98

Ticking the Boxes..100

The Project...100

The Team 102

The Market 104

The Industry 106

The Exit 109

Chapter 6 – Getting Traction **111**

The Business Model 112

Ownership Structure 114

Exit in Mind 115

The Right Team 115

The Right Amount of Funding 117

Focus on the Cash 118

Risk/Reward...119

Timing and Sequencing 121

Setting the Plan 121

Set Achievable Milestones..122

Allocate, Measure, Report 123

The Action Plan 124

Measurement...126

Project Status Review 127

Commercialisation Point..127

Traps to Be Aware Of 129

Groupthink...129

Internal competition 130

Hoarding..131

Valuation..131

Chapter 7 – Value from Licensing.................................. **133**

Managing Risk 136

Proof of Concept...138

Royalty Rates 139

Licensing Performance 140

Timing 141

    Savings and Shortcuts..142

Revoking Licenses 143

Future Products..143

Chapter 8 – Crowdfunding....................................... **145**

Why Crowdfunding?..146

Do Your Research 146

    Quirky 147

    Crowdfunder...148

    Appbackr..148

    AngelList..148

    Indiegogo...148

Your Target Audience 150

Planning the Event..150

    Corral the Hungry Herd 151

    Social Media...152

Your Compelling Exchange 154

The Video 155

After the Campaign..156

Chapter 9 – Scaling Up **159**

Target Market Focus...160

Your Competition 161

Path to Cash Flow 163

Forming Strategic Alliance Partnerships........................163

Group Networking...164

    Individual Networking 170

Scaling Your Networking..172

Parallel and Gray Marketing...................................173

   Giant Client Syndrome 174

Scaling up 179

   Where One Plus One Equals Twenty.................180

What Makes Venture Capitalists Different? 182

   Preparing to Scale Up......................................186

Chapter 10 – The Value Realisation Point..................... 189

Valuation for Trade Sale...192

Pricing...193

Perspective...194

Paperwork..194

The Players...195

Payment Options 196

Profile of IP buyers..197

   For Trade Sale...200

   For Joint Ventures 202

   For Listing...203

   Aggregation...205

   Licensing...205

   The Dash for Cash 206

Chapter 11 – Owning the Process 209

The Cornerstone Investor..211

Becoming the IP Magnet..211

   Where Do You Fit?...212

   The Filtration Process....................................213

Aggregation..216

   Start with the End in Mind 217

   Vague Higher Authority 217

IP Selection 218

   Build Your Innovation Cell............................219

Project Status Review 220
The Walking Wounded 220
Separating Ownership 221
Periodic Value Testing 222
From Magnet to Magnate - Spinning Out IP 222
Aggregation for an Exit 223
Maintaining momentum 223
Fairness 224

Chapter 12 – Value from Expertise **227**
My Story 229
The Monetized Membership Model 231
The Ascension Model..233
Lead Generation..237

Appendix 1 – A Pre-Venture Capital Checklist......................... **239**
Your Pitch Deck 240
Cash...241
The Company...242
The Market...243
The Team 243
The Operations 244
The Finances 244
Their Internal Questions 244
Summary..245

Appendix 2 – Network Pre-Meeting Checklist Form............... **247**

DEDICATION

This book is dedicated to all of those inventors who took a deep breath and jumped off the end of the job line, with nothing but an idea and a suitcase full of passion. Those of us, who thought we could all be walking around with computers in our pockets twenty years ago, would have been marginalised.

How many people in the 1980s would have thought to deliver 1,000 of our favourites songs in that same computer and let us use it to speak with each other? This wasn't a big idea – it was simply a shift in an existing idea with a sound commercial model wrapped around it.

We all have ideas every day. We just need a method of unlocking the value in the business model that can make these ideas valuable.

Here's to the people who see differences and stop to ask: *"How can I make this into something big......?"*

INTRODUCTION

So you have an idea that you believe could set the World on fire. Your friends and family tell you it's a winner, but you are not sure what to do about it. People around you have been helpful, looking at the industry and suggesting that if you got 2% of the global market you will be a squillionaire. Some may even suggest you should quit your job and convert the garage to a research facility/workshop and start on your dream tomorrow.

These suggestions might all be helpful, but not realistic. If you want to move your project forward, through the traps and roadblocks of the commercialisation process, you first have to examine what you have, very objectively. To help with this, I would suggest that if your "winner" project eventually loses money, it will be your money that is lost, so be careful and practical about how you view this opportunity.

Most academic text books are written from the Corporate or University perspective. If you have an idea, but do not have the unlimited marketing resource of a research department, no access to research and development facilities or funding, or even the backing of patent attorneys, those processes may have very little relevance to you.

The examples they use, such as Steve Jobs with Apple, are prominent because they're the most unusual stand-out wins, which did not follow the regular process, but won anyway. To follow the Apple process for commercialisation is the same as to follow Steve Jobs as an innovator. There will only ever be one Steve Jobs and in our lifetime, this provable likelihood of getting another innovation firm like Apple, in the cycle it was in, is highly unlikely.

Marketing books sometimes touch on innovation and intellectual property, but they again handle these from companies like Coca-Cola and other multinational entities, which have little or no relevance to the average inventor. There are many textbooks and Ph.D. theses, which focused around the needs of large corporate and tertiary institutions, for the commercialisation of intellectual property. This book does not provide a solution for those types of organisations.

At the other end of the scale, are innovators who are focussed around their idea and have no business tools to make the idea into a reality. Some watch TV innovation shows and think investors will fund their ideas. Although this is good entertainment, it's not informative and it does not depict the common reality for the commercialisation process for intellectual property. It presents you with one winner and several losers, in a 30 minute episode, which is about as far from reality as you can get.

The premise of this book is that there is an inventor in all of us. Some ideas are good, some not-so-good. Others might be outstanding, but the only ideas that should be enacted upon, are the ideas that can be developed and owned, that people or companies will pay a reasonable amount of money for, to either make or save them money or time.

It could have been many years ago, that you scratched out an idea, perhaps on a napkin at a restaurant, when having lunch with a few friends. You believed the idea had merit, but you didn't have the time or other resources to do anything about it. Perhaps you threw it in a drawer, in that old desk in your man-cave (garage) at home and forgot about it. Perhaps you picked this book up and have read to here, because you feel that idea might actually be worth something. It might, but what stands between you and a hungry herd of people willing to pay money to buy it or use it, is a process.

This book is written for the amateur inventor who understands that the idea he or she has right now could be valuable or worthless, but unless they get a better understanding of the process of commercialisation, they might never be able to tell the difference. Sure, we can rely on others, such as Patent Attorneys and Accountants to give us an independent opinion, but in most cases, their opinions may be skewed towards them being able to bill fees for helping you, if you were lead to believe the idea has merit. This book is a guide to turn your scratchings into something of significant value - to create that billion dollar napkin.

The good news for any reader is that you do not have to have a tertiary education to understand and implement the techniques we give you in this process. Most of what you read in this book is the combination of more than 28 years of commercialisation, from an idea to a public company and everything in between. There are no magic formulae and the stories I give are not stand out "against all odds" wins, but rather techniques and processes to help any innovator to progress his or her project through the maze of commercialisation.

This book is a sensible step-by-step guide to commercialise your innovation, from an experienced advisor perspective. As a management consultant, specialising in the commercialisation of intellectual property for more than 28 years, I have seen my fair share of projects succeed and fail. In fact, I have seen mediocre projects succeed and amazing projects fail, which has lead me to provide this work so you, the reader, can be better equipped to achieve your planned outcomes with your own innovation.

In mid-twenty09, I was invited to conduct a business diagnostic review with a group of investors (including the inventor) of a very promising electrical engineering product. We have a structured half-day workshop scheduled, in which we extract everything we need to prepare an independent report on the innovation and then we follow this up over the following weeks with external research on the industry and the markets.

From the first minute, I could tell the project was doomed. This was at the "introduce the team" stage of the workshop, where I am being introduced to the people, their roles, their backgrounds and their expectations. The first to speak was the inventor, who was also the majority shareholder. In under a minute, he had already convinced me that he was not prepared to share his secrets with my team or his own investors, he would not comply with any proof of concept or trade sale audit requirements, and required a substantial amount of money for the project, to be redirected to him for his past work.

Just to support my first minute's assessment, he went on to spend nearly 50 minutes telling me how good he was, how good is project was, how good his other projects were and finally how poor his competition was. This was a person who had no intention of letting go of the command and control process that makes businesses work. He was a self-taught mechanical engineer, who clearly had an innovative brain and was capable of great things.

However, it was obvious to everyone, including the investors gathered there, that this meeting was going to go nowhere, and the project would probably follow. At the first morning tea break, I stepped aside with the principal investor (who was only a minor shareholder) who pleaded with me to keep going, when I expressed my concerns directly to him. He was clearly trapped and desperate, so I spent the rest of my presentation testing for a process that could resurrect the project.

The project failed for one reason. When we look at projects, as management consultants, we use the same 5-point criteria that venture capitalists and most Cornerstone Investors would use. These are the people, the product, the industry, the market and the people. With every innovation project, it starts and ends with the people. If something goes wrong with an intellectual property project, it is almost always the people. Most venture capital investors will choose the people over the product every time; because they know they can source other intellectual property to give to the right team, which will then ensure success.

This project did not proceed, as none of the investors who were already in the project, including family members of the inventor himself, would continue to fund it to the next stage. Sadly, the world will have to make do without his technology, but this may be developed after his demise, by his family. This is not an uncommon story, and will crop up time and again through this book, as the people element of any project will always be the most vulnerable point. Although this project suffered from an abundance of ego and was blindsided from competitive capabilities, the technology itself could have been superb, but it's earning as much as it deserves to today. Nothing. The ego and ignorance in that project could be calculated as risk and subsequently killed off the project within minutes. This is not an isolated a story.

Another interesting point of this electrical product story is that the project team believed they could be ready to list the company and raise money on the stock exchange, to bring the product into production. They felt they were ready for international distribution and licensing, but the assessment we conducted that morning suggested they were at the idea stage and not even at a concept stage, despite them raising and spending a considerable amount of seed capital.

It is important to understand at what stage your project is, within the life-cycle of the commercialisation process. If you are realistic about this first aspect, then your planning is focused on the transition from this stage to the next stage. This risk reduces your scope of attention and will focus your allocation of resources to achieve your next objective, rather than focusing on making an idea an international company, well before it's ready.

Regardless of where your project is in the process, I maintain that you start this book from the very beginning and not jump to the sections most relevant to your understanding of commercialisation and your project as you see it. In some cases you may have got where you are because you skipped a stage or two and in many cases, that could cost you money.

In management consulting terms we refer to this as BLOT, which is business left on the table. If you felt your project was more akin to the activities of chapter 6 for instance, you would almost certainly start at chapter 1, but only if someone was to convince you that the difference could be up to 50% of your value by the time you reach chapter 12.

With my decades of experience in intellectual property commercialisation, I can tell you this is the case. If you jump ahead, you miss the foundations of what the process is and you may fall victim to people who recognise that you do not understand the process fully.

ABOUT THE AUTHOR

Daniel O'Connor is a rising star of the international business world, a highly sought after Intellectual Property commercialisation specialist, a Management Consultant, a Business Coach, a Company Director, a business growth expert and author. He is a member of the UN Taskforce for Innovation and Competitiveness and contributes regularly to international membership learning websites.

Daniel's skills are in demand across Australia and internationally in a range of specialist areas. He has spent 28 years in professional practice, specialising in Intellectual Property (IP) commercialisation – helping innovators to get their intellectual property into the market and achieve massive success.

Currently the Consultant Principal at Incub8IP, Daniel has helped many companies to incredible international trading success with their concepts. He is also a Founding member of the Formula1forBusiness franchise, which helps service professionals to double their net income in ten months, using a safe and structured set of incremental changes that makes the changes "stick." He has assembled a team of highly skilled consultants so his business can service the needs of a huge range of clients. Daniel has planned and managed public and private companies over his career with outstanding results.

A constant thirst for learning has seen Daniel study in Australia and internationally to acquire many qualifications including Bachelor of Business (Marketing) and an MBA in International Business, as well as being on his way to completing his PhD in International Business (Intellectual Property).

Daniel has helped boost companies to success through his knowledge, advice and mentoring not just in relation to their

IP, but also strategic management, acquisitions, performance management, business systems and process improvements, business and industry aggregations, public listings, interpersonal coaching and change management. He has established new business ventures in engineering, e-commerce, knowledge management, automotive, agriculture and aquaculture, pharmacology, food, tourism, microbiology, electrical and electronics as well as IT and education industries.

Daniel relaunched Success Motivation International (SMI) across Australia in twenty06 after an 18 year absence. SMI is a franchised organisation of independent distributors dedicated to personal development through a specific goal setting process. Daniel worked with senior managers, business owners and executives with leadership and goal-setting programs to help them exponentially increase their success.

Throughout his career, Daniel has established firms in China and other emerging markets, as well as negotiating and establishing distribution networks for Australian firms throughout South East Asia. Daniel's extensive knowledge and network of contacts internationally means he can open doors to opportunities that others may not be able to access. He has enabled clients to raise millions of dollars in grants and investments as well as commercialisation funds.

Daniel has worked and studied extensively in several countries over many years. His international experience includes countries in Asia and South East Asia as well as the USA and Canada. He is a member, fellow or associate of numerous organisations, including the Australian Institute of Management, the Australian Institute of Company Directors, and the Australian Marketing Institute just to name a few.

You can follow his innovation blogs and publications on www.makemyinnovationhappen.com, www.profitfrompatents.com and www.billion-dollar-napkin.com. His main training website is www.incub8IP.com

WHAT INDUSTRY LEADERS
ARE SAYING:

"This is the most insightful, impacting & important book you will ever read, for building value from your patents. It is a blueprint for success for every inventor and is full of ideas, wisdom and strategies that can change you are thinking forever".

Rajen Manicka PhD
BioScientist and multiple patent-holder
CEO of Galen BioMedical USA and Holista Colltech Limited.

"Billion Dollar Napkin is for the entrepreneur who is not flirting around the edges trying to grow a business with simply a hope and a dream. This is a serious, nuts and bolts blueprint to building AT LEAST a million dollar business - if you fail miserably and have to settle for a lower sum.

The bit that makes this read fascinating and worth your weight in gold is that it's all from experience in Daniel's track record of building 7, 8, and 9 figure companies all over the planet that had once started out as scribbled nonsense. This guy has the freakish ability to take a simple concept and engineer it to the point of getting it to IPO level and I can not recommend this read with any more urgency for anyone serious about building a real business from a solid idea.

Buy it, study it, and put it into action play by play and you'll find success in your venture".

Josh Smith
Advertising Entrepreneur
Sanford & Smith - San Francisco, USA.

"Done with a clear understanding of the real world, answering the questions that you have when you try to convert your idea to a business. This book is useful, clear and focused to solve key problems you will not see coming. I found this an excellent tool that will keep the process moving.

Daniel's worldwide expertise, advising hundreds of corporations and successful entrepreneurs, provides a clear understanding of the business rules, making the complicated simple and makes this a must-have for your learning library. This is the road map to value and will help every entrepreneur to present their projects how the capital providers need to see them".

Jose Escudero Roldán
Managing Partner
BMI Capital Partners - Spain

"Billion Dollar Napkin takes the often mystifying intellectual property establishment and monetization process and packages it up for layman or expert consumption alike. Daniel O'Connor populates the clear Step 1-Step 2-Step 3 format with perfectly clarifying (and entertaining) anecdotes that lay it all out. If you have IP to exploit, or think you might, this book will certainly save you some time and money."

Eric Illowsky
Chief Operating Officer
Inter-American Management LLC

"We often hear words like disruptive innovation, game changer, and entrepreneurship at University or in the media. What we do not learn is the real process in making intellectual property a reality. Billion Dollar Napkin demystifies the process of what it takes to make an idea into a

business. Anyone reading this book, even if they do not have an idea, can learn what role they can play in the innovation process. This book should be given to every University student."

Chris Shulha
Doney Leahy Private Assets

CHAPTER 1

Myths Models & Rules

"Your work is going to fill a large part of your life and the only way to be truly satisfied is to do what you believe is great work. And the only way to do great work is to love what you do."

—Steve Jobs

Understanding the Rules

To begin this basic explanation of commercialising your innovation, I would like to share a few golden rules with you that if you take anything away from this time you spend reading this, it should be these points. They are critical and fundamental to your achieving the first outcome you are seeking, which is to obtain the required funding to transition your project from an R&D project, to a commercial project. These fundamental rules should be used to measure your transition program and your capital-raising activities, to ensure you have a substantially higher probability of success.

Get Realistic

Before you prepare your project to present to parties you believe may be interested in investing, you have to get realistic about your expectations. From my 25 years in commercialising innovation (from within a University environment to corporate and one-man innovation teams) I have found that the novice principal innovator generally has the same five unrealistic expectations. These are:

- A central-city building with their name on it
- A million dollars - in their bank - not the company's
- 51% of the shares - after distribution
- Chairmanship and full control of the board (usually populated with their friends)
- A global "fact-finding tour" (holiday) for the principal(s) and any immediate family

Working backwards from this level of expectation can be daunting and in some cases, impossible. I have reviewed a very promising project from the South West of Western Australia which met 3 of my 4 critical criteria. However, the innovators' expectations were so unrealistic that we passed on the project (to the dismay of the other shareholders!).

In most cases, innovators anticipate an immediate global demand for what they have and possess very little access to research on demand, pricing, costing and competitive offers. The days of the "better mousetrap" have long gone, due mostly to the increased tooling and start-up costs for any commercial venture. It sometimes comes as a shock to innovators that products they will compete with are delivered at wholesale for around 18-22% of their eventual retail price.

Emotional Commitment

In more than just a few of these cases, the principal is not likely to be movable on some or all the above points, for a separate reason. I understand that in many cases, the innovator has dedicated years of his or her time to the project and in some cases, much of the family savings. When I am brought in and present a realistic expectation of commercialisation, it can devalue their commitment made to the project.

For instance, if the principal has dedicated 6 full-time years to the project and considers his time to be worth $20 per hour, there are very few (if any) R&D projects that can internalise that level of value expectation. When I walk in and present a value expectation based on the net present value of future returns, it can devalue the project to a level that renders my opinion dangerous.

Build the Business Model First

Have you ever watched Shark Tank or any of these other investor pitch shows and wondered why some of the great ideas don't get picked up? If you see something that you think would be an exciting opportunity, are you surprised when no one on the panel makes an offer?

It doesn't surprise me that most of these innovations don't get investors. If you think about it, there is hardly ever enough information for any investor to make a decision, but the interesting thing to recognize is that the questions from the panel members are generally about the business model itself - not the product.

In most cases, astute IP investors are always looking at the business model before they look at the product. They know better than to fall in love with the innovation itself and focus mostly on how the ideas could earn money to pay them back and how it would continue to grow their investment. The intellectual property doesn't need to be outrageously different or elegant in appearance or function. It simply needs to have a killer business model which will provide substantial and growing return for the investors and the principal.

Some of the most profitable ideas are not complex product-based innovation, but rather just a unique service delivery which can increase the earning capacity of those who adopt it. A product or concept does not need to be patentable to be attractive to investors. The patent is simply a method of ensuring that your competitive advantage can remain yours in the future.

In the early 1920s, the race was on to develop the next best mousetrap to combat the growing number of rodents in large cities. More and more wild ideas and complex mouse catching contraptions were developed, but when it came down to it, very few of them sold as well is the single disposable piece of timber with a spring- loaded snap trap on it. The reason why this remained the biggest-seller during this era and well beyond, was because it was disposable.

Once the trap had caught a single mouse or rat, its job was complete and it was disposed-of, complete with the rodent. The business model was *"you never have to clean this trap - it catches a rat and you throw it out."* The elegance of this business model was that it met the single greatest demand for most housewives at the time, which was *"I do not have to handle anything with a dead rodent stuck to it."*

The lesson for patent-holders and other intellectual property owners is to focus on the business model as it relates to the users – focusing on their specific requirements. Once you understand how the buyers would use your solution to solve their existing problem, you can calculate what they would pay for that based on their current cost to solve the problem.

Only when you can confirm that you can solve this problem more effectively, at a lower price and using an acceptable process, can you prove the demand for your solution is going to be better than the existing or other solutions.

If the business model supports this, your opportunity has value. If it is elegant and patented, it could be worth even more. However, with the elegance and patent protection, it might be worth nothing, if you don't have an attractive business model to work with.

So where next you see an episode of shark tank or any other investor pitch shows for innovation, look closely at their presentations from a business model perspective. You will be asking yourself how any of the panel could possibly make a decision about investing in any of the pitches, based on what is presented.

The answer is most likely to be in the experience of the panel members, who can look at the innovation and build a model inside their heads. Although they might accurately calculate the value of the opportunity based on future returns under their model, it is highly unlikely that they will share that with the presenters, as this would reduce the risk I the opportunity and increase the value – which might cost the investor more.

The lesson here for all innovators is to calculate the true value of your product from the business model and not from how sexy or elegant your product or service is. Astute investors don't get hooked on the product or service. They will jump in to a goods business model, which shows how the company will accelerate and how they will receive a return on their investment.

It may be hard to accept at first read, but let me break the most important news to you early. The most critical aspect of your innovation is not the product or service – it is the business model. If you later stand in front of prospective investors and ask them to invest in your product, you are most likely to fail. If you ask them to invest in your business model, which has your product as its foundation but has a clear method of making money, has identifiable demand from a sizable and approachable group of buyers and has a reasonable life cycle for profitability, then you will more than likely get the investment you need.

You may have watched innovation pitch shows on television where people with ideas ask a panel of investors to individually or collectively jump in. This might be great television, but it is not a reality. Almost all the inventors are spruiking their projects and in five entertaining minutes the panel have to decide if they want to be part of the business when all they have seen is a description of the product. This makes any decision naive or premature at best.

When recruiting for resources (partners, funding, joint-ventures, etc.) we have to firstly pitch the business model, then the benefits of this to the party you are pitching to, before you then start on what the product is about. If you keep this first rule paramount in your thinking, you are more likely to be able to become a commercialisation machine, as you will be able to overcome the hardest roadblock of the entire process, the first independent funding stage.

Project Focus

The next golden rule is to "fish only where the fish are." It sounds crazy right? But follow this logic. Most projects have iterations or applications that can expand the project appeal to different markets and ultimately yield a greater return. However, if you tackle these opportunities all at once, it will require far greater resources (less equity for you) and ultimately slow your project down. In the business diagnostic review we conduct with every new project, we list, prioritise and value each opportunity and then "park" all but one or two, which we then transition into early-stage commercialisation. We liken this process to standing in a corridor and facing 20-50 doors (opportunities). We then drive the selection process (un-emotively) to close all bar one or 2 of these doors, to ensure the project can generate early revenues to justify the high-growth required to survive, as well as to fund the parked project at a later date.

I recently engaged with a very innovative technology team and although they had previously raised and spent millions, they were not generating cash flow on any of their opportunities. After the business diagnostic review, we parked 15 of the 17 opportunities and focussed all of their resources on one joint-venture with a large global technology player and one niche application for one of their secondary products, which was going to generate immediate and significant returns. After struggling for over a year without investment commitment, we were able to demonstrate a clear path to market and got the investor confidence we needed to fund the project forward.

Rule of Costs

My third golden rule is three-fold. It can be stated that every project will cost more than you expected, take longer than your predicted and yield less than you forecast. Even the most experienced developers and commercialisation teams will struggle to meet or exceed their stated commercial objectives. In some cases, when projects are rendered un-workable, the investors apply a (rule-of-thumb) discount for risk. I cannot think off-hand (from more than 400 projects) of any single project that was delivered to market on or under time and on or under budget.

This leads us to the next rule, which although terribly obvious, is rarely addressed as it should be. In all of your dealings and your projections, you must be fair and honest. It is true that if you are to pitch your product alongside other projects, to the same select investors, there is a high probability that some of the other innovators have exaggerated their value, returns and development maturity as well as hidden some of their faults or issues. This may give them an immediate competitive advantage to secure funding over you, but they will crash and burn eventually and in most cases, astute investors will never put more money into a project if the projections were not realistic in the first place.

I have been involved with several projects where the innovators have exaggerated their product readiness, or hidden their technical issues, and who justify it (later when the project is dissected and revealed) by declaring that the world should not be deprived of their project and they didn't lie for personal gain, but rather for the "greater good" (of the people having access to their widget or their employees continuing to get paid).

Offer Commercial Products Only

The next golden rule to remember is that very few investors will actually or intentionally invest in R&D. There is far too much risk to invest in projects which are still being developed, but more importantly there are many innovators with the R&D mindset that cannot get comfortable transitioning to early-stage commercialisation. As a result, they tend to switch to one of their many other projects or start to develop other iterations or applications for the project, in order to remain in their comfort zone.

One clear distinction in most projects is that R&D does not yield a return, but commercialisation does. Innovators must be prepared to transition all or part of their project into the commercialisation process and park their additional opportunities, until the first project generates sufficient cash to feed the parked opportunities.

Stepping Away

The next golden rule is to be prepared to let go. If your expertise is in developing technical projects, you need to accept early that there are going to be other skill sets required to commercialise your emerging opportunities. The prospective investors may demand that you have another team for the commercialisation and that they may insist on doing it a very different way than you would. If you insist on control of the project in a business area you know little or nothing about, you can expect the project will falter. Astute investors will recognise this and will focus on projects where the developer has passed the project to a separate commercialisation team, with the appropriate skills, experience and perhaps contacts, to make it happen faster.

A very significant percentage of innovation projects fail because of the innovator's refusal to relinquish control. Within months or even days, the novelty of having a positive bank balance in the project's accounts becomes less of a priority than their perception that people are running away with their asset (which in most cases, is now jointly owned).

I have found that the best way to overcome this is to establish an independent panel of experts to advise and guide the project, so that key personnel, who understand the technology but have no business acumen, can still contribute with relevance. We tend to refer to these as Technical Advisory Panels (not Boards – there are legal and liability issues associated with Boards) and we generally structure these as voluntary, with a promise of options to be granted for length of tenure, at the pre-IPO stage of the commercialisation cycle.

Packaging Your Innovation

In many cases I have reviewed, the innovation is wrapped up in an overly-complex cluster of technology. Sometimes this technology is not completely relevant to the commercial components of the innovation, but may have been intricately bundled just so that the innovator and his innovation remain prominent and relevant. Skilled business diagnostic reviews should un-wind the cluster of technologies and innovations and quarantine all the commercial components before transitioning these individually, to the early-stage commercialisation process.

Spin-Out

Several years ago I was invited to meet with a group of highly innovative developers, who had produced a multi-language computer program, so that software developers who did not speak English as their first language, could still develop fluently in any one of 27 languages and then (with one keystroke) flip it into English, or any of the other languages. This enabled programmers with less than a highly-competent level of English, to understand the finite differences in string commands. For example, the word REVERSE can have at least six acceptable functions, including turning around, to flip, to invert, to move backwards, etc.

For anyone with less than a very high level of English, the confusion is immediate, but not that evident. When this word is used to manipulate a column or stack of numbers in a sequence, it can become buried in hundreds of lines of code and never understood. This project has substantial merit, but started out with no business model to commercialise it and was so large it would be impossible for most investors to see its potential.

The first phase of early-stage commercialisation was to break the product down to a tangible market application, which was to have it accepted into the Education System in China, for the standard programming language taught in high schools and Universities – in Chinese – using a full character set keyboard.

The politics of integration (into an education system) in China has proven so overwhelming for these developers that some five years later the project still did not have traction. So we broke it out to other applications, We elected to focus on the project as the tools, to build significantly faster programs (this used approximately 5% of the code required for most programming) and we set about identifying some of the developments we could create that would have an immediate, profitable market application.

The first three products are to be released commercially in 2016, and will be funded in their own right, as separate spin-out projects. In summary, while we could not commercialise the core innovation immediately, we were able to use it as a tool to develop very innovative products and then commercialise these, as well as to use these spin-out projects to showcase the power of the software as a tool.

Spin-In

Most projects reach a point where additional resources are required to get them to a positive cash flow position (where they are earning more than what is being spent). This is usually the time developers start to look for government grants or investment funds. An alternative is to conduct a "make-or-buy" analysis and see if it is not cheaper to source some other technology and integrate it into what you have, to overcome the issues or roadblocks you have.

There are millions of "one-product" projects out there that will find it almost impossible to raise capital, although the concepts may be sound and the products almost ready to go. These projects are better off aggregating under a strong development or commercialisation leadership, to share the market development costs and to present a bundled group of like products to prospective distributors, for a complete package. I discuss the opportunity of aggregation in a later chapter.

Presenting what investors want to buy

In most cases, the pitch presentation to prospective investors is done by the principal innovator. There are some exceptions but, for the most part, I would argue that this is not best course of action for highly sophisticated products.

Firstly, the more complex the product is the more beneficial it will be to get someone independent to present the project (not the product). Secondly, if you get somebody independent (perhaps the marketing manager – not the technical manager) he will be able to make the presentation around the market demand, the costing and pricing, the products of services this will replace and all the relevant marketing information with which the investor will start.

Innovators will need to remember that it is very rarely about the product, but more about the things it can do, for whom, and how it can save time and/or money. Investors will also need to be presented with the future iterations, so that the project is not perceived as a one-product company.

Standards or regulatory requirements may need to be met before a project commercialisation process. Some industries (such as the therapeutic goods industry) have stringent requirements which can take years to obtain the approvals to package and market your products.

These standards approval requirements can take years and if your product does not perform as stated, you may have many more years of lab trials, field trials and final approvals. You may also then need to repeat these approvals tests for different jurisdictions. Some tests can be simple, such as the test used for children's toys. The minimum size part for any child toy for the "under five years old" age-group must be of a size that does not fit into a plastic film canister (which are not readily available anymore). This standard made sense twenty years ago but is very hard to explain to the average teenager.

Prove the Concept

One of the absolute requirements for capital-raising is to be able to provide an independently validated performance test of the product, such that it can prove the functionality and all the claims made. This must be conducted by an independent authority and can be expensive if you intend to use a university department to conduct the validation.

The Proof of Concept stage is critical, but can vary for different products and services. In some it is simply a matter of independently proving the product meets or exceeds all the claims made in the offer document. I have a current project which required proof that it could capture carbon from exhaust flues, of a coal-fired power station. We determined that most Universities were a little too keen to see the project and so we had the Proof of Concept performed by a State Government chemical emissions and pollution control centre.

With their background in testing power station exhausts, they encouraged us to also test for nitrates and sulphates, as well as other particulates, which proved very useful later. The project was able to claim 99.6% effectiveness in capturing carbons, in an exhaust pumping at 140kph. Because this test was performed by a Government Testing Authority, it was never questioned.

Another recent project in the automotive industry required Proof of Concept validation, before investors releasing additional capital and the principal investor releasing a substantial order for his own large fleet of mining vehicles. Because the primary market was to be the USA trucking industry, we elected to conduct the testing at a testing facility in America, operated by a global industry standards authority.

Distance can play a major part in the validation program, as this project was to discover. The engine (sourced locally – second-hand) disintegrated during testing and the project did not attain Proof of Concept. The principals decided to have it tested by fleet owners and drivers in the USA, who all were suitably impressed, but their results were not independent or scientific enough to encourage investors to continue the funding. The net result after several months of planning and two rounds of testing was an invalid Proof of Concept and a stalled project.

Other Factors

In this book, I have deliberately not addressed three core areas – IP Protection, Offer Documentation and Legal Agreements. These are specific to your jurisdiction and should be managed by a competent and experienced Patent Attorney, IP lawyer, Contract Lawyer and/or an accountant with Corporations Law experience.

Targeting the Investment Profile

When most developers and innovators look for funding, they are essentially seeking people like them, who will share their vision and understand the reward expectation. In most cases, this profile of investor does not exist. Unfortunately, as the realisation dawns on the developer team, they start to widen out their hunt for investors and they focus on everybody who will talk to them and present their project as:

- A great opportunity to make money, and
- A great new technology that will achieve wide recognition.

Unfortunately, people become interested in new products or technology for a myriad of other reasons. This reason becomes their currency, and in order to attract the most likely to respond investors, we have to understand their currencies and present the project framed around that. This is primarily why large-scale presentations to groups of investors should be avoided.

In order to target the right investor profile, we need to firstly understand who would best suit this? We need to experience they can make more than just the expected return from being involved, which might include some measure of exclusivity within their local geographic market segment, if they are a potential user.

As developers preparing a customised pitch for each target group, we need to understand fully what their primary currencies are. Depending on the size of the investment organisation, you may have several parties you are pitching to and each may have different currencies.

It is almost always easier to get a Company to invest than an individual because in the case of a sophisticated investor, you have to get past the investment advisor, the broker, the accountant and in some cases, the private banker. This additional process will introduce delays and the need to re-sell the same opportunity to all of these advisors or stake-holders, before you can proceed. In some circumstances, it is inconceivable that an advisor will encourage early stage commercialisation projects, particularly if they are not remunerated and also given that that portion of the investment funds is then "out of circulation" for the advisor who would otherwise be able to transact these otherwise liquid funds several times during your proposed investment period.

Reframe Your Pitch

In most cases, the information documentation and the presentation material is all about the product and perhaps the returns. The investor has other priorities, which are usually focussed around the list provided in the next section for you. However, the how and when of getting out, is paramount in any investment strategy. There is nothing worse than a minority shareholding in a private company, which the directors have decided to change their plans and all the shareholders are frozen in the investment until the company is either sold or listed.

One of these key issues for investors is the exit method. Most experienced investors have a clear preference for how they would like to exit and will generally hunt for opportunities that fit their profile. A little homework before your presentation will enable you to structure the deal to provide their preferred exit. This may mean your project has to modify its plans, as once you lock in that investor you are committed to that option. Some of the most popular exit options include Packaging for Trade Sale, Public Listing, Licensing, Distribution, or Franchising (Growing-it-out).

Investors are going to want to see who the team is and they will also need to see a commercialisation-focused team taking over from the R&D team. One of the most fundamental tenets in the intellectual property process is that the skills required to complete the R&D are radically different from those of commercialisation. If the R&D team does not accept this, the project will almost always descend into conflict and chaos and most astute investors will know this. When Edison employed teams of researchers and product managers to work in teams, this single fact made him his millions.

These teams were kept separate (still within Menlo Park facilities) and were instructed not to liaise with each other except through formal chaired meetings. There was always conflict and many brilliant people resigned in frustration, but the Centre was very productive and given those days' value of money, would surely be outperforming Silicon Valley for returns.

Investors will also want to see when you expect to realise your return on this project. For their safety, they will need to see the innovators exit after the investors but also, in high-end value projects, that some of the equity provision for the innovators is held back and allocated against future performance milestones.

As most R&D projects are not making a profit, there is little use for tax concessions, unless they can be cashed out. It might be prudent to take your investment partner on as a Collaborative Research partner, where they pay the commercialisation expenses as a partner and can then claim the R&D expense through their own entity.

When considering how to slice the pie, it may be prudent to look not just at equity, but at debt funding or a hybrid (convertible note or C-note) where some or all the funds is provided as a loan (with perhaps a fixed and floating charge over the assets of the company - including the IP) and an option to convert all or part of that if certain milestones are met within a certain period.

There are also many ways to sweeten the deal for commercial or corporate investors, including access to the product (particularly if they are a vertical market partner) or to exclude their competitors for an agreed period. I recently had a project where a very large firm paid a substantial (six-figure) sum to be able to access the product when completed, for a five year period with initial two-year exclusivity in their particular demographic. This fee was paid into a joint account, exclusively for approved commercialisation expenses.

What Most Investors Will Want

There are eight core elements of an attractive investment for corporate or commercial investors. To deliver the most impact in your presentation, you need to direct your pitch to cover all of these and focus on the one you know (through pre-presentation interviews) is the "hot button" for this group. There are rarely more than eight and there is usually only one that has more emphasis for any specific investor group. The most popular I have dealt with in the past 25 years can be summarised as:

- Clear evidence that what they provide will cover everything you need to turn a profit

- How they will get their return, the size of it and when it is expected
- How the investment funds will be held and allocated
- How you have assessed and intend to manage commercial risk
- What returns are expected for the company as well as the investor
- What is the hidden value for them (leverage, reciprocation, exclusive benefit, etc.)
- Who are the people who will control the project and also the funds
- Independent Proof of Concept

How to ask for more and get it

Most developers do not consider the sensitive touch-points in a capital-raising negotiation. Generally the most sensitive issue is "you want how much cash? …..Now?" but if a deal is properly structured, it does not have to bite into the company budget, which will generally not have catered for contingencies like this. Some of the softer pitches can also get you more value, based on factors that may seem insignificant to you but substantial to the investor. These can include:

Little or no cash upfront: I was involved in a recent project where a substantial proportion of the project was offered to a high-profile business person, for no cash up-front. However, when the Proof of Concept was delivered (at the trade sale audit) the investor was to place an order for this product, for his main business. This order was to be in the magnitude of around ten million dollars, which would provide the company immediate working capital and a blue-chip client reference.

Scrip for scrip rollover: Not all investment pitches are for cash. As mentioned (above) the offer can be increased if the cash cost is reduced. There may also be some tax relief in your jurisdiction, which you should explore with your accountant. One critical issue with acquisition of any intellectual property being exchanged for scrip is the progression of payments as the IP is completed and commercialised.

There is always inherent risk in making sure IP performs as it should and market acceptance is as predicted. Investors need to have a retention or protection from faults and failures. In the case of IP purchasing, we generally structure a retention payment after the product has operated flawlessly in the company or in the field, for a period.

When shares (in the investor company) are involved, these are not able to be taken back upon performance failure, so most deals will structure the issue of equity against the performance milestones for IP acquisition. If investors had agreed to issue shares for the acquisition of IP, they will generally undertake to provide parcels of these shares when particular agreed milestones have been achieved. It is not a common practice to provide these shares in a single upfront delivery. Shares, once issued in the vendor's name, are the property of the vendor and could immediately afford that entity voting rights.

Performance based reward: If the IP has been developed with little or no cash from the innovator(s) there is a propensity by most investors to discount the value and to rethink the risk and commitment associated with the innovators.

This could make asking for a fair share a little difficult, unless you make the request conditional. This is best done with the formation of a third party company from which equity is allocated to the innovators (usually at $0.000001c per share) with an immediate uplift when the investment funds are placed into the company and shares (equal to the amount of equity – usually at 1c or 10c) is allocated.

If your proposal includes an additional allocation for the innovators and/or the developers, subject to agreed performance milestones being met, then you can realise a greater reward without additional risk to the project or the investors.

Stages and Phases

This book is a tool to help you transition your project through each stage of the commercialisation process. We refer to the transitions between stages as phases. For brevity, I will describe the stages but focus on the phases to help innovators transition from stage to stage. In my experience, there is always one stage where team members are most comfortable, but lingering too long in any given stage can kill a project slowly. The focus of this book is to explain that although we may be more comfortable at different stages and be hesitant to transition, in the interest of commercialisation we must push ourselves out of our comfort zones.

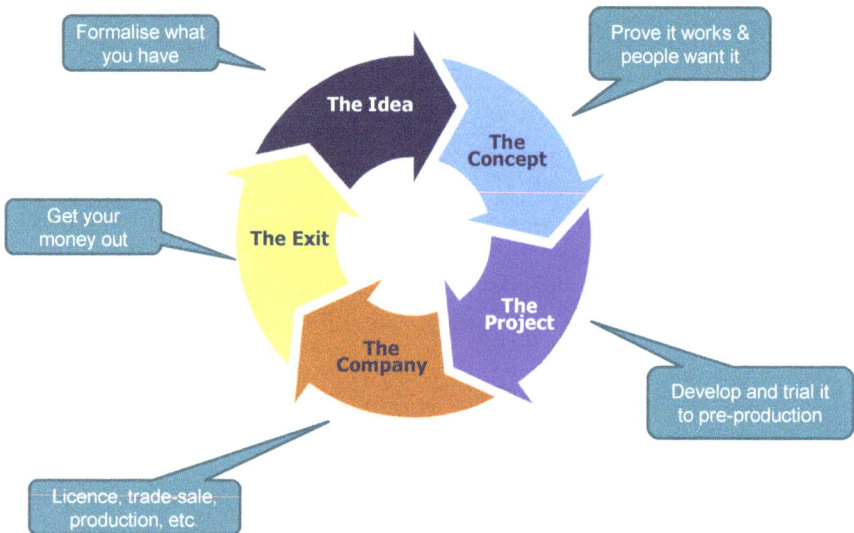

The Idea

To confirm the project is in Stage 1, an innovator must test the idea for ownership and protection, uniqueness, utility, potential unit cost, and sale value to see if what is being considered can actually generate a sustainable income over its life cycle. The most critical part of this is ensuring that the idea can repay the investment and effort made by the innovators over the development period. When we examine Stage 1 projects, we talk of getting realistic. Some of the cold, un-emotive questions must be examined to ensure the idea can become something and ultimately be successful.

The first question in the Stage 1 evaluation is "what do we have?" To answer this, you may need independent validation of market demand (industry reports on the buying/spending patterns of your target market), as well as both a detailed description of the idea, much like a provisional patent application, as well as a six-word "elevator-pitch" description.

It might be prudent to use caution in deciding whom to ask for advice on determining value for your idea. Friends may not want to disappoint you, and patent attorneys might give you a positive impression so you will be motivated to take the next step and patent your idea. This could be fatal at the idea stage. In almost every circumstance, an independent market validation should be sought before any patent advice.

If you plan to develop your idea into a concept, you first have to fund your research and then do the development. You may be able to do the product research on your own, but you cannot do the market and industry research because you will not be able to assess your idea objectively. You may have heard stories of triumph where passion and stubbornness paid off and a billion dollar project was born, but this is the exception.

There is probably no statistic to support this, but the amount of provisional patents registered around the world in contrast to the amount of global conversions of these provisional tenders suggests that a lot of ideas do not make it. At the same time, plenty of inferior products are successful because their innovators examined the market objectively and delivered what the market needed and wasn't able to get. One of the lessons of funding early R&D is that your most likely target for investors is usually friends and family. Considering the risks, early investors are probably investing in you and your reputation and not necessarily your idea. There is a simple explanation of why people cannot get public financing for an idea: it does not yet have value.

For friends and family members who invest, you still must pitch more than just your idea to them so they have confidence in the project. You should be up front about how much you need for the project to be profitable so they (and you) can calculate how much time and cash this will take to make money. They will want to see that independent people have looked at the market demand and given an objective assessment. Independent research from sources like IbisWorld can support your assessment of the competition and how the industry tends to react to new players.

The Concept

The first step in the transition from an idea to a concept is Proof of Concept, an independent test of your technology or innovation that proves that it performs as required to the standard required by the market. Part of the Proof of Concept is the availability of components to build the product, as well as the ownership of every component including software. This also includes the capacity of someone to build it for a price. If your project is a retail product, you should operate on a cost-to-retail ratio of seven to one to ensure that you and others in your supply chain are profitable.

If you have an engineering product that you will sell direct to the end customers, your variable cost per unit is an inclusive maximum 30% of your selling price. As a general rule of thumb, a project must contribute one-third fixed cost, one-third variable cost, and one-third contribution to profits. Fixed costs include administration, marketing, and other set costs of the business.

Variable costs are only costs associated with each new product that you build. The contribution is what's left after you've paid your fixed and variable costs. This should always be a minimum of one third of the selling price, multiplied by volume.

Once you have your Proof of Concept, the project team faces one more hurdle: the trade sale audit. This is sometimes referred to as the internal trade sale audit because there is no actual trade sale involved. This does not change the process, which is important as a handover of all product concept ideas, knowledge, and experience from the R&D team to the commercialisation team. These teams might include the same company personnel or might be a completely different group, but the trade sale audit must be handled as formally as possible.

The process of a trade sale audit involves assembly of all the product derivatives, prototypes, the list of materials and suppliers, sales and supplier contracts, operations manuals for the product, packaging, and other associated goods and services or rights.

The process used to formally hand these objects across the table from the R&D team to the commercialisation team ensures that the R&D team understands that they no longer are developing the product. At that point, any additional expenditure will be handled by the commercialisation team.

Experienced investors will scrutinise how the trade sale audit was performed to ensure that those who developed the project have willingly handed over all responsibility for the commercialisation. There are few researchers and developers who can successfully commercialise a project or product. When given commercialisation responsibilities, an efficient R&D team will listen to the market but then go back to R&D to modify the product and create iterations that satisfy market niches they may find. This prevents the product from ever making it into the commercial arena, as it continually cycles through R&D to become bigger, better, and brighter.

The Project

Once you have performed and documented the Proof of Concept and trade sale audit, you are ready to transition the product from the concept phase to the project phase. The concept phase usually involves two key steps. The first is to raise sufficient capital to turn the idea into a profitable product with established markets that continually grow or that larger competitors will want to buy from you.

The second step is the commercialisation process, which may involve a full grow-out program of offices or outlets (perhaps franchises) in each of the target market regions, although this is the most expensive option. However, it represents the best value for a trade sale to a large national or international company once you can provide an established track record of sales and established distribution networks.

More commonly, people search for distributors, manufacturing representatives, reseller networks, or other established channels that service the target market. I recently worked with an electronics communications company, which was established as the Australian reseller for a North American firm. The company had developed a special box that did something their market wanted. To distribute globally, they simply approached all international resellers from the North American principal.

Although still in its early stages, this project has two trade sale opportunities: with the principal firm they represent and with that firm's major global competitor.

Having your first sale or contract—even if it is subject to approval—will go a long way to lessen the risk associated with investment. This can sometimes be more than just a sale for new products. In the last five years, I worked with a property management firm that wanted to launch a cloud-based commercial property system for larger retail and commercial property owners.

Their best objective was to secure an anchor client by selecting one of the big five in the industry in their country and offering that firm a 30% share in their project, if they could manage their property portfolio through this portal. With their first "sale" providing several millions in revenue, they were able to launch the project into the next level of its growth phase and start hunting for additional clients.

This is not an uncommon occurrence. I worked with an engineering group on a product for the transportation industry that had provable savings on operating costs. They secured an anchor client (a mining giant) by offering them 30% of the project in return for ten million dollars in pre-sales once the product was approved. Even venture capital companies will take notice when you have millions in pre-sales from a blue-chip company on the books.

The commercialisation process will vary significantly depending on the type of product or process, the target market, and the price and type of product. It may be food, medicine, or engineering components, which require radically different commercialisation models. One of the first elements of a commercialisation process is approvals. Nearly all products need some type of approval to be sold, consumed, or used, and these approval requirements may vary across jurisdictions.

Although the project phase generally will earn money for the owners, you will still be expected to complete a trial to determine that the needs of the buyers are being met at the right price. In most cases the project would have only prototypes, but you may choose to build preproduction prototypes in order to ensure economies of scale for high volumes.

Any capital raised will need to cover the cost of production prototypes, early production runs, patents, and patent conversions, as well as conducting field trials, which can be documented for technical and marketing purposes at a later time.

Early funding should be used to validate market demand in core regions, as well as examine the degree of substitutability of the product to other market segments and other products to this market segment. This will help manage risk in the future.

In order to fund these needs, which represent the early part of the project phase, we encourage Cornerstone Investors to participate. A Cornerstone Investor differs from any other investor inasmuch as he places a substantial amount of capital into the project with an expectation of securing a substantial share of the project and a competitive advantage for any product he uses. A good Cornerstone Investor is one who could use the product in a market segment not focused on by the project, so that a specific geographic or demographic licence agreement can be issued as an incentive. There may be investment incentives for the Cornerstone Investor based around a substantial tax deduction for investments made.

This is usually done under a collaborative research partnership, which would enable the investor to participate actively in the development phase by paying the bills of the early-stage commercialisation, as and when they become due, as a collaborative research partner. There is a high probability that the development project will not require any tax incentives in the first few years, whereas a Cornerstone Investor may be able to benefit directly and immediately, thereby reducing his cash exposure to the commercialisation risk.

With the backing of a substantial entity such as your Cornerstone Investor, the project is better positioned to apply for and obtain commercialisation grants, which are matching funds to a limit of

$100,000, or in some cases $500,000, to complete particular tasks in the commercialisation process. Sometimes these funds can be applied to the salaries of key executives, provided that the project can prove the benefit of having these people on board. This effectively doubles the investment outcome for the Cornerstone Investor.

Some of the commercialisation activities that may be eligible for grant funding include packaging, operations manuals, early discussions with potential distributors or resellers, and early production quantities to prove sales demand and margins.

Once you have the production and growth program in place, start to build your track record. When you have a trading history and you have cash coming in—and growing—each month, you should use that information to update your funding documents for expansion funding. This is the ideal stage to transition the project to a company.

The Company

Transitioning a project to a company is all about expanding the operation by accelerating the production and distribution, as well as expanding into other regions. This expansion funding will be required to convert and perhaps later defend the patents the project holds, as well as providing an opportunity for the principals to acquire vertical markets through strategic acquisition, building a brand and propelling the company toward profitability.

Expansion funding must deliver the project to a cash-flow positive position, the principal method of mitigating risk associated with the commercialisation. In some cases you may need additional funding for materials and manufacturing equipment, but in many cases this kind of cash need may be debt financed, against orders obtained in the previous stage.

The company phase is where many projects fall down. Running a company that produces 50 products a week and sells them profitably might make some comfortable. However, if the project can be crafted to produce 5,000 products a week and all of these can be sold to willing buyers, a completely different skill set may be required.

Many projects get caught in the company phase because they start to get comfortable. Or, the partners do not agree on how to move forward. The faster you can transition through this phase, the more value you can obtain at the exit stage.

Most people are looking for the blue sky in any company investment. They will look at how sales are tracking to determine if the growth trajectory is realistic. As you accelerate, everyone will want to be your friend. Most of your risk is depleting, and your commercialisation tasks are becoming more focused on patent protection, product iterations, and improvement for production and application.

When you reach a satisfactory level of sales and a firm indication of more with the right funding, you are ready for your final transition to your value realisation point, or exit.

The Exit

Although most venture capitalists call this the exit, it is more of a value realisation point than an act of removing yourself from the product or the project from your company. Many projects are sold off by VC investors or developers at a set value realisation point, and there are many projects that spin out from larger companies after being incubated through the first four stages of development. However, there are just as many innovation projects that are retained by the developers to enhance their current product range or to complement the company in attracting new markets.

The value realisation point is generally referred to as an exit because it is about the investor getting the option to realise his investment. Back at the transition from concept to project, you may have invited a Cornerstone Investor to provide the transition funding to your company phase. This Cornerstone Investor should have a structured exit that may include cash with interest, or he can swap for equity in the company. The critical aspect of this is the Cornerstone Investor must have a set structured exit or value realisation point at his investment stage, and the commercialisation team must work closely with him to ensure he can exercise his right.

There are many options to best maximise the value at exit. The five most common exit valuation processes are the trade sale, an industry aggregation, a public offer, a reverse takeover of an existing public company, or the grow-it-out model of expansion. These are explained in greater detail in other chapters.

The Phases

Most projects will gravitate to one of the stages, which is usually where the principal feels the most comfortable. It makes sense for a project which the principal is happiest in an R&D role, could tackle a marketing problem with an R&D solution. This would rarely be the perfect solution and could increase the project burn rate with the potential to stall the project in the longer term.

In order to better understand transitioning, the most significant difficulty is the work that needs to be done, to move a project from one stage to the next. This transitioning work is generally referred to (by consultants) as the phases. As most of the difficulty in IP Project progression is in these transitions, I have provided a list of the phases that link each 2 stages, so that developers and innovators can understand the requirements.

What differs across most investors is timing. Some investors become involved in the very early stage. Others enter a project after it already has a track record of sales with a provable growth program that adding new dollars and contacts will exponentially grow the project to a new level. Let's discuss the four most common transitions between the stages of commercialisation.

From Idea to Concept

Transitioning your project from an idea to a concept will generally require less funds and more effort. At this point the project has little value to the market and cannot be presented in detail except through secrecy agreements. The most common investment profile for this phase of a project is the principals' friends and family members. In some cases friends and family may pay for provisional patents or meet expenses as the inventor transitions from part-time to full-time in the project. Usually, there is little need for investment other than for basic equipment or supplies. This suits the friends and family model, as it's less formal and requires less financial accountability and paperwork for the innovator.

From Concept to Project

This funding is commonly referred to as seed capital, but in innovator terms it can be represented as a single source Cornerstone Investor. Cornerstone Investors commonly have an interest in the industry and/or the market, an understanding of the project and some passion toward its outcome, and an opportunity to obtain a benefit for his other operations once the technology is available. He usually has contacts in the industry and/or the market to facilitate early-stage trials. In Australia, we generally seek a minimum of $500,000 from the Cornerstone Investor, which places that investor in a sophisticated category with the expectation that they are responsible for their own due diligence going forward. This is known as the gold card rule under Section 708 of the Corporations Act.

From Project to Company

The next stage is venture capital, and there are strict unpublished rules about VC selection criteria that can preclude particular projects and projects at particular stages of development. Venture capitalists provide resources to start-ups, but they also define start-ups as companies that are already in operation with good sales, an excellent track record, and high growth potential. Venture capital companies are not interested in investing in technology. They like to invest in companies that are profitable and that have exclusive access to technology to support their growth.

A venture capital investor will conduct a full audit of the company, including an IP audit, review all finances, and verify the current assets and liabilities of the company. Most innovators consider venture capitalists as people who will invest in new projects or project start-ups, but they are generally focused on projects that already provide a reasonable return on investment and show substantial growth or growth potential.

Venture capitalists generally contribute more than just money. In most cases they specialise in particular industries and will provide introductions to market players for the innovators. This will generally accelerate growth and in some cases will justify the fees they charge in their speed to grow the project. Venture capital companies often like to be represented on the board and will carefully scrutinise other board members for suitability. There is no room for friends and family or old colleagues at the boardroom table of a venture capital investment.

From Company to Exit

The final investment category is not required for every project. In most cases it makes more sense to trade sale the opportunity to a large publicly listed company in exchange for shares in their organisation. Other options can be a trade sale for cash, but that option requires funding for the projects that you intend to list on the public stock exchange. This funding is called pre-IPO.

If a venture capital company has been involved at a previous phase, they are likely to control this phase carefully. The details provided here are for companies that do not use venture capital but believe their project will grow to a point where they need a sensible exit, and an initial public offer of the shares on a public stock exchange appears to be the most appropriate. There are many capital companies that provide this service, mostly affiliated with or connected to legal and/or accounting firms.

This is because the majority of the funding that is required will be spent on legal and accounting services and reports. Another option for the IPO is a reverse takeover of an existing listed company that needs to restructure their operations because of product failures or lack of cash. In some of these cases the IPO firms will take control of the company through the deed of company arrangement and will restructure the shareholding to a more manageable level that reflects the actual value of the company. They will then seek an opportunity to place the company under another company in exchange for what is sometimes a majority shareholding in the project.

This essentially creates a publicly listed company (usually with a name change) and a bunch of legacy shareholders with a minority share whose core investment was in whatever the company did before. Provided that the number of shareholders meets the relisting requirements of the stock exchanges, these projects can be funded by the IPO companies in exchange for equity and re-listed on the stock exchange. Although this may sound like an attractive option, there are many traps. Innovators are advised to seek professional advice before selecting these options.

In this chapter, we have examined the process of early-stage commercialisation for the average innovation. I have deliberately avoided references to large corporations, pharmacy developers and manufacturers, soft drink companies, and multinational software companies, as I believe these have been handled adequately in University textbooks.

This chapter provides a structured method of transition for your project from an idea to a fully functioning and profitable company. I hope it helps you focus your expertise, research and development, and early-stage commercialisation on your next big idea.

It is rare for researchers and project developers to become good at commercialisation, and it is also rare for commercialisation experts to become good at managing and operating corporations. There will always be exceptions, but at the end of the day if we love what we do, this ultimately helps us make far more because we are motivated to build the best product we can create.

CHAPTER 2

The Six Roadblocks

"If you want to see the single most likely thing to make you fail in business, get a mirror..."

—Daniel J. O'Connor

There are six common roadblocks that almost every commercialisation project will encounter. If you understand these roadblocks, it will be easy for you to prepare your project so that you can navigate past these when they turn up. They are usually in the same order and almost every project will hit each roadblock in various forms of severity. When a commercialisation project dies it will usually die at one of these roadblocks.

Understanding the key roadblocks and preparing a project to overcome them is going to increase your commercialisation chances by at least double. In some cases, innovators do not understand the roadblock that is stopping and are focused on something entirely different whilst whatever they do, almost inevitably fails to get past where their project has stalled. From knowing what you have and do not have to presenting it in the most appropriate way (structure and ownership) are equally as critical as having a planned exit which will deliver an achievement outcome for you and investors (exit options). Having a sound understanding of these roadblocks and how they can very quickly stall a good project, can help you steer away from the dangers associated with these.

To best take advantage of this knowledge, you will need to firstly assess where your project is and then ensure that you have ticked off all the key points in roadblocks that are before yours, to ensure that you are in fact at the roadblock you think you are. In many cases people approach prospective investors well before they have the structure and ownership worked out and wonder why the projects aren't funded. If you can tick off all the issues associated with structure and ownership and proof of concept, then the roadblock of funding will not seem as insurmountable as it has.

Structure & Ownership

Structure and ownership is the first major roadblock that most commercialisation projects will hit. In most cases the owners believe their first roadblock to be funding but in fact they do not get their structure right and haven't sorted out the ownership pre-and post-investment, there will rarely be investment.

Getting Out: When packaging the project for funding, most innovators do not focus on how the structure will look in the future, from the shareholders' (including the investors and their own) equity. They do not consider that investors will at some point, wish to realise their value and take their money out of the project. Most successful investment proposals will provide a value realisation point and an estimated value at that point, so that prospective investors can calculate their true return over time.

Valuation: The project should have a fixed, independently prepared valuation, which will help to justify your claim to the equity you will retain. This should not be a statement from your accountant suggesting that the thousands of hours you have dedicated to the project, should be multiplied by a reasonable hourly rate. Your claim to value should be based on the ownership of the IP and the net present value of expected earnings, discounted for risk.

Ownership: One significant underlying fear for every investor at any presentation is that the inventors do not have full ownership of the intellectual property. Because the development history is absolutely defined independently, the investors must rely on the innovator to ensure that the patents cover everything to do with the intellectual property and that there are no collaborators who can make a claim on the project or the intellectual property, at a later date.

Borrowings: Another issue that may not be disclosed at the pitch is the amount of debt in the company. Some of this could be informal debt from the family of the inventors and may not be recorded. Any latency claims to the intellectual property that could surface during the commercialisation process, may directly affect the investment value of all the owners, including the Cornerstone Investor.

In most cases the value presented will be in the concept, the ownership of the concept and the net present value of future earnings. If there is contribution by owners in time or cash, this contribution is recognised as commitment to the project and should be rewarded with consideration in equity.

If there are any funds placed in there specifically to acquire items such as patents or legal fees, these funds cannot be considered loans to the company by the principles or others associated with the project. It is important that the structure reflects that all of these borrowings have been converted to equity and that the company will have no debt going forward, before they present an offer to prospective investors.

Past Collaborations: By the time a project is ready to ask external parties for funding, there have generally been many changes in the design or R&D team. Some of these parties may have exceeded by choice and others may have been negotiated out of the project. The expectation by any prospective investor will be that you have insured any party that has withdrawn from the project in favourable or poor terms, will be subject to some sort of written agreement on the ownership of intellectual property and/or the promise of equity from within your proportion.

In some cases, early contributors can stall a project completely because they believe their contribution is pivotal to the success of any product and there value interpretation becomes unreasonable. Cornerstone Investors are looking for this type of situation in any project, so they can avoid the inevitable fights later.

The most common outcome with these structural and ownership issues, is unusually long delays in reaching each of the milestones, as all parties can become distracted by emotions like frustration, greed and anger.

Value of Past Input: one key point to remember in structuring your value in a project is that your value is calculated on the future, not on the past. Any amount of labour that you've contributed to a project does not warrant any reward. If that labour results in a new project which can create an income stream in the future, the value is calculated on that income stream and not on the effort to date. This is why the most experienced developers, will test market their new concept before they develop it, so they can calculate what sort of net present value of future earnings their project can generate.

Exit for Investors: Most project innovators will calculate a fair value for investors, in proportion to the contribution they are willing to make. One of the key issues overlooked for most investors, is the timing and value at exit. This exit does not have to be a selling back or selling off of the shares equity provided for the funding. It is more likely to be a value realisation point at a certain decision milestone in the future, where the investors can recognise their value independently through the listing of the shares or a trade sale offer to a third party.

Having a defined exit will enable investors to calculate the time value of their money from investment to exit, whereby providing a set calculable return on investment. In our formula1 For Business program we offer a Cornerstone Investor the opportunity to fund an aggregation of businesses which we collect together for listing on the Australian Stock Exchange.

The aggregation process takes between eight and ten months to complete and the Cornerstone Investor funds are used for acquisition of the shell and the aggregation process itself. We are able to provide the Cornerstone Investor with a commitment to deliver $1.2 million worth of equity in the listed shell at relisting, for an investment of $500,000 into the project as the cornerstone investment.

Because this investment period is less than twelve months, the return on investment is more than 140% when annualised. With this type of stated return, the prospective investor is generally satisfied with the return and is then only focused on the risk. If in our presentation we prove we have covered the risk adequately, we will have no problem in encouraging prospective investors to place their $500,000 into the project for the required ten months.

Ownership: There could be a possibility that English property could be co-owned by other parties who could have collaborated on the research and development or contributed ideas, equipment or expertise in the development phase. Care must be taken to disclose and document all potential ownership issues, so that these can be overcome well before the value can be realised. It would be prudent for innovators to negotiate with any third party and obtain signatures of clearance under a freedom to operate mandate. Obtaining signatures from third parties for limited or no claim on the innovation will be a critical step in the packaging of the IP for commercialisation and funding.

Patents: If there is an opportunity to patent the idea or innovation, then this should be explored with a professional patent attorney, as quickly as possible. It is essential that any disclosure to 3rd parties is covered by a deed of confidentiality or nondisclosure agreement, so that ownership and novelty can be claimed and there has been no public disclosure before lodgement of a provisional patent application.

Most innovators tend to register the patent in their own name and there is no problem in doing this. However, in most cases the technology will have to be vended into the commercial entity before the investment funds are received and this will include the provisional patent. Assignment of the provisional patent and all rights associated with it will ensure the company has unfettered rights to commercialise and in return, the company will be responsible for maintaining the conversion to a full patent at the end of the provisional patent period.

R&D Team versus Commercialisation Team: The more proficient your research and development team is, the less likely they are to be as proficient in commercialisation. Part of the structuring process involves separating R&D from commercialisation and recognising that the skill sets required are radically different. Most successful projects are managed by commercialisation teams who understand they have no expertise in research or development and have R&D teams who recognise they have no expertise in commercialisation.

This issue becomes important when projects transition from ideas to project and start the commercialisation process. As you commercialise a project the intellectual value shifts from technology knowledge and know-how, to marketing manufacturing and distribution know-how.

Proof of Concept

This roadblock is by far the easiest to overcome and yet is the most common. Inexperienced innovators fail to realise the importance of an independent validation of claims, which can satisfy prospective investors or collaborators that you have something of interest. Some of the elements of a proper proof of concept would include the following:

Test Criteria: Planning out the technical test criteria is an important aspect of proof of concept. There has to be validating statement that says "if this can do that, for this price, in that time, then it must be deemed to work" there may be many other proof of concept criteria for individual specific projects but this is where the process should begin.

Soundness of the Business Model: There are many innovative technologies that fail the business model test. I've seen projects which have impressed me to the extent where I can accept that they are game changes for the way people do things, but these have not made money because their commercial business model is less sound. The technology must have a sound business model to wrap around, to become commercial. The key onus on any IP commercialisation project is getting people to pay money for using or buying the project or product or service. They must immediately understand how it will make or save the money or time.

Supporting Scientific Theory: When planning out a proof of concept program, you would need secondary research on scientific theories, which support your project or which further identify the problem you are attending to solve.

Secondary Market Research: The proof of concept process requires that you independently validate how many willing buyers are represented in the market demographic you are targeting. You need to see independent proof that they exist, they have the same problem that you can solve, they have the means to pay for your solution and they are frustrated enough to take action if the right solution is put in front of them.

It is equally as important for the testing of pricing and costing associated with the end product. Investors and collaborators need to know that a sufficient number of the target market have indicated they would pay sufficient money for this product or service if delivered soon, and that the product or service can be provided at a substantially lesser price than what people are prepared to pay.

Internal Trade Sale Audit: One of the critical components of a proof of concept process is the internal trade sale audit. This is generally conducted around the same time and requires an independent technical authority (such as a patent attorney or IP lawyer) who can prepare and check a list of claims, items, features, supporting documentation, et cetera, that will support the completion of the R&D process. The internal trade sale audit is not in fact tied directly to a trade sale, but is used within the company to transfer a project from R&D to commercialisation.

This process will incorporate the act of transfer associated with all the items on claims relating to the project. If an R&D team cannot produce or do not produce everything at the internal trade sale audit, then the project is either not ready for commercialisation or the developers are holding back information. Either way, the project is not ready to have money spent on it at that time.

Freedom to Operate: Immediately before launching new product lines, a sensible precaution would be to have an independent third-party research the industry and market, to determine if there are any new products or technologies that may directly compete against what you have. This process may also rigorously examine other patents to determine if these patents could prevent the company from trading or could result in litigation. For high cost developments and high cost launch projects, such as pharmaceutical and medical products, a freedom to operate check is almost a standard procedure.

Funding

Funding is the most common roadblock experienced by nearly all innovation teams. In some cases the innovators have few or no prospective investors to put the project in front of and in other cases they are presenting their pitch to many parties but not getting any interest. There are several key reasons why funding becomes a problem for commercialisation projects and the following list is only a few of these.

Not Ready: In many cases, projects are presented for capital-raising before the developers have been able to build sufficient value to justify the asking investment value. Projects should be able to be independently valued by an auditor or similar party, to justify the amount of equity to be granted to the developers and a further valuation of the value that can be achieved, when funding is provided to meet the next stage needs.

It is always good to get an idea of what the value will be from prospective investors, but unless the innovators can establish their value and compare this to the cash component being provided by investors, there is really no justification for ownership or equity in contribution.

Ownership of the IP: Before you can ask investors to join you in a development or commercialisation project, you need to be able to prove that you own the intellectual property associated with the innovation. This includes providing statements and/or evidence that no other party has or can make a claim to the IP from this investment point.

Some innovators tend to keep their IP registered in their name and give a licensed to a company in which they would like to raise the capital. If the company that is to commercialise the intellectual property does not have full and unfettered rights to the project in perpetuity, it does not have ownership. Very few investors will be interested in project in which ownership is not included.

Most innovators appreciate that without vending in the intellectual property, they have very little to offer the commercialisation process, in comparison to hard cash. An acceptable alternative could be that the intellectual property is vended to the intermediary who will receive both the IP and the cash, but should that company get into any difficulties the IP can revert back to the original registrar for a menial amount such as one dollar.

Not structured correctly: Care must be taken to ensure that the project structure will accommodate the needs of investors' partners and owners beyond its current position. Investors will require a return on investment, but they will also seek to be protected from any liabilities associated with past indiscretions of the directors inside or outside of the project. This suggests that incorporating a new legal entity and vending the intellectual property into this company in exchange for equity, could be the best solution for everybody. When the investor provides the funding, they are simply purchasing shares in registered company, which already owns the intellectual property and all rights associated with this.

There can be no question on ownership of the intellectual property and the application of the funds within the company. On the other hand, when funding is provided in tranches it is also important for the IP owner to understand that the associated equity will should also be provided in an equivalent fashion so that investors do not have an inordinate say in the corporate direction before they have completed their commitment.

No Independent Validation: Most of the issues associated with Independent validation of claims have been discussed in proof of concept above. There are times when I am asked to assess intellectual property for third-party investment and I generally have a first informal meeting with the IP owners where I list their claims and then provide them, usually at a later date, with at least and a summary table of sources of Independent validation which I asked them to complete. In some cases the intellectual property owners tell me they do not have access to the validation required and I am then in a position to report to the prospective investors that the project is not investment ready.

No Assessment of Risk: Like the Independent validation requirements for technical claims, some intellectual property owners fail to validate their positions on risk. Having a table of risks and associated mitigation processes, together with Independent validation of these is the most appropriate appendix required for any funding presentation.

Little or No Assessment of Competitors: As with risk, people fail to properly assess their competitors. If you are small company in a small industry and your competitors a small company owned by major corporations, those competitors are going to have radically different responses to any new market introductions you might plan. An entirely different marketing model is required to tackle a monopoly or duopoly than would be required for a homogenous industry such as a retail outlet.

Market Estimates are Guessed: I have seen many pictures for finance in my 28 years in this industry. In that time I have seen many good and many bad but the worse pitches other great pitches that provide guesstimates of market access or penetration, based on a percentage of the overall global market. When I see statistics such as "1.5% of the global market equals" I involuntarily shudder and audibly groan, as I know the entire pitch has started its spiral and is about to crash land.

People suggest that they can access these millions of people through the Internet but the cold hard reality is that most Internet viewers are not immediately attracted to whatever your claim would be and some who are looking for your service might not have your message resonate with them at that time. If you placed a link to a Facebook page which promotes your message of personal self-improvement, you could likely argue that there is very few people on this planet would not wish to improve their current situation. The reality would be that you would get less than half of one percent of people even read your ad and for those who click through, you would still get a less than 1% of these seeking further information.

This would mean that you could pay hundreds of thousands of dollars to get hundreds of people who have clicked through your advert, read your website and have responded by completing a questionnaire or application form. This still does not mean to say that any of these viewers will pay you money. Those hundreds of thousands of people are not representative of the 1% of the global population and so promoting percentages of global English speaking people who "wish to improve their current lives" as a demographic segment, would just lose your project all credibility immediately. To provide a context, I can suggest that there are more millionaires in US dollars in India, than there are people in Australia and Singapore combined. Although this is a geographically tight target market, accessing them through the noise of 900 million people would prove problematic and expensive.

Cash for Debt: In some cases, the original pitch for money provides little detail on the allocation of those funds when received. I have seen cases where a greater percentage of the funding allocation was to be used to repay the inventor some of the cash or time that he has committed to the project. Asking investors to pay for the innovator's share of the project will kill the deal faster than asking for a donation.

If you are the inventor and you've made a commitment to build this project, part of your commitment has been your time and your money to get the project to a point where it has some substance of value. The investor now stands before you and is offering to buy a share of the combined value of that company. If your investment and effort has generated an independent valuation of $1 million and the investor is being asked to provide $500,000 then the combined value of that company is $1.5 million and the investor should be entitled to 33%.

His funding is for 33% of the value of the company and does not allow the investor to pull out some of that money to repay his out-of-pocket expenses, without reducing the overall value of the company and thereby increasing the value of the investors share.

IP Owners will Not Provide Enough Equity: Some of the disappointments in writings on intellectual property commercialisation are the examples of some significant wins such as Steve Jobs with Apple. The reality is that a fixed value is given by an independent authority at or near the time of asking for investment and the combined value of the company would include that intellectual property and the funds.

The value of each parcel of equity can readily be calculated based on this formula and in most cases investors would not be interested in insignificant parcels of equity unless enough of those parcels are in independent hands. In most cases, I would advise investors not to enter into early-stage commercialisation collaborations with innovators who do not provide a minimum of 33% at investment or who do not provide a seat on the board or other technical advisory panel involvement, as an option.

No skin in the game: Many investors will look at the commitment by the innovator and measure this by the amount of money that they have put into the project. Money hurts more than effort and perceived value of financial contribution at early stages is far higher than if the innovator had no direct cash investment committed to the project.

Everything has to be Cash/now: Two of the most common misconceptions in capital-raising are that all the equity has to be for cash and that all of it needs to be provided upfront. I have structured deals whereby equity was offered for a ten million dollars sale of the product at first production and I have been involved in others where accommodation and access to scientific computers and equipment were traded for equity in the project. In many cases, cash is not needed upfront but has requirements to meet milestones at certain intervals.

Where cash is to be contributed incrementally, there is risk that the contributor fails to meet their commitment along the way. This could have a detrimental effect on the company development and there should be provision in the equity agreement that a penalty could result. I have recently experienced a project where the investor ran out of funds after the first 20% and the innovators were lucky enough to have a penalty clause which allowed them to buy back the equity for a nominal amount and allow the investor to write off his research and development as a tax loss.

Focus

The most common roadblock in the focus category is the focus on the product at the expense of the market. Ultimately, any product or services is designed to solve a problem, fill a need or scratch an itch in some way. Before a product (solution to the problem) is even contemplated, a meticulous examination of the need and the prospective buyers (including their propensity to pay for a solution they do not have right now) will shape the solution product or service.

It is vitally important to continually revisit this market problem and check that there are no shifts or developments in the need or available solutions, which could affect what you do along the development and commercialisation way. This does not suggest you should focus on variants or iterations of your product to suit emerging opportunities, but rather to ensure you are nailing the need in the biggest demand area, so your commercial product will have a chance to be received well.

A focus on other projects can fatally distract most innovators. When an R&D team has more than one project it will become very difficult for them to put projects on hold while they service the ideas that receive the funding. This leads to R&D teams working on unfunded projects whilst being paid for from funding received in other projects. When this behaviour becomes exposed, the investors may have recourse to lay claim against other innovations based on the evidence that there funding is being used to develop those projects.

Distribution & Scale-up

One of the more common non-financial roadblocks is logistics. Some projects have excellent preproduction prototypes and have methods of building the product in low quantities, but have failed to address the issues associated with scaling up. Some projects -particularly in metal fabrication with large components - can prove too problematic to use a production line process and therefore do not lend themselves to economies of scale. This will restrict the amount of sales that can be delivered and therefore reduce the company's growth to the capacity provided in the current facilities.

Part of the business plan is to scale the project to produce the current products or services to a wider group of buyers and in greater numbers. Where most information memoranda or business plans address the issue of marketing, many fail to address the potential on economies of scale or the problems associated with scale of production.

No Distribution Channel: As extension from marketing guesswork in the financial proposals there are many projects which do not have established distribution channels for local national or global markets. A business plan may address this issue by simply saying that when funding permits, the company will expand into other markets, but this effectively dismisses any growth projections that are provided.

At the very least, the project must have a plan for scaling up production, distributing the finished product, packaging and displaying the product and at least evidence of some dialogue with prospective distributors or retailers.

Packaging and Shipping: One of the questions that a venture capitalist will ask a project when they're ready to scale is how much their packaging and shipping is going to cost for individual products. They want to know that the innovators have addressed this issue in the commercialisation planning and if they haven't, there would be very little prospect that the project is going to get funding. This is the same with Cornerstone and Seed Investors, who look to make sure the project managers have addressed all the minor issues as much as they may have addressed the major issues.

These types of qualifying questions will immediately alert a Cornerstone Investor to the depth of planning that has gone into the project, beyond the research and development. Part of the advantages a project would enjoy with having a separate commercialisation team, is that they probably possess experience and know-how around the scale, packaging, distribution and marketing of the product or service the project has.

The New Skill-Set; Having extensive experience and knowledge in the research, development, prototyping and build for any particular product or service, does not guarantee that the project team has experience in commercialisation. If the commercialisation is to be performed by the R&D team, most investors will recognise the two major issues in this dilemma.

The first is that the R&D team is going to have to learn how to do commercialisation and they'll be learning that while they spend the investors' money.

The second key issue is that when marketing responses and commentary are received by the project, the propensity for developers is to take the product back into the R&D cycle and generate another iteration of the product to satisfy that market issue. This inevitably prevents them from failure in commercialisation, as their product is back into the R&D process and cannot be judged as complete.

What this situation can do, is deplete the allocation of commercialisation funding and render the product a more-capable research and development project but a less-capable revenue generating product or service.

Exit Options

Before the preparation of any information document, you need to be able to answer several deep and meaningful questions, including how and when any investor and even you can exit the project. Having a clearly defined entry point with cash amount required and when spells out what you need. On the other side of those negotiations, an investor is going to need a clearly defined exit point with an estimate of the cash that he will be able to take with him at that time. We call this an exit option but in practice it's more of a value realisation point.

At that point there is no obligation for the investor or the innovator to sell their product or project. However, having a valuation and opportunity to sell will make all the difference to investors who may otherwise feel trapped.

The most common exit options include a targeted trade sale, an initial public offer of shares into a public stock exchange, a reverse takeover of an existing listed shell which can then be relisted onto a public stock exchange and finally, a transformation to a grow out model which may include staff equity, a management buyout or even a franchise model. These value realisation points do not force any party to sell but will provide an ideal opportunity to sell all or some as well as to give an independent valuation point for new buyers.

CHAPTER 3

The Seven Deal-Killers

"If you want to be understood, you must first seek to understand"

—Stephen R Covey

There are hundreds of issues that could slow a project down, but in my experience there are seven core deal-killers that almost every project will experience. In cases where these deal-killers are managed or avoided, it is through good planning or sheer luck. Experienced project developers will address these issues before they become major problems. Be aware, even with great planning these deal-killers are capable of depleting funds and time or derailing a project entirely. These seven deal-killers are described below, in no particular order.

Rushing to Patent

You identified a problem people or companies are having, that is constantly costing them money. You have a great idea that you know will solve this. You know everybody will want to buy this and how much they would pay. You need to now turn the idea into a product and the product into a Company.

Where do you direct your limited funds first? There are 6 areas in which you need to spend money and in most projects, you won't have enough for all of these. In no specific order, these are:

1. Patenting
2. Market validation research
3. R&D to Proof-of-Concept
4. Design, tooling and packaging
5. Building up Inventories
6. Negotiating (funding, marketing, distribution, trade-sale, licensing, etc)

At first glance, these all appear to be equally important, but which one should have priority? Nobody can adequately afford to pay for all of these up-front. Certainly, you could make a solid argument for any of these to take priority over others, and that is exactly what the providers of these services will suggest to you.

You see, the designers, patent attorneys, consultants and toolmakers all know that you will not be able to afford to do any more than one of these, so they will design their pitch to convince you that you need to spend all of your reserves in only one area and that area should be them.

It makes sense, right? If you walk away from a meeting with a design engineer, the patent attorney will get your money, if he sells his story better. They all know this and they pitch really hard to make you decide you need to spend all of your money with them. And the successful ones are really good at pitching, more than actually doing the work.

Not me. I will tell you where you should NOT spend your money, simply by providing you with a process that shows you how to schedule each of the services and only activate each feature when these become a need, instead of a want.

I have nearly 7,000 inventors who have linked with me on LinkedIn over the past 2 years. In most cases, these inventors have been convinced they should invest their limited funds into patent protection – BEFORE THEY KNOW THE VALUE OF THE DEMAND. It makes sense to protect it before it becomes valuable, right? But what if it turns out to not have any value?

Let's face it. If your solution costs $100 to make and it saves the buyer $20, what is the point of patenting it? Also, if this idea will solve a $100 problem and will cost $5 to make, you have to know HOW MANY PEOPLE HAVE THE PROBLEM AND HOW AFFORDABLE IS THIS GOING TO BE FOR THESE PEOPLE WITH THE PROBLEM? If there are fewer than 500,000 buyers in a year, how will you pay for 5 years of patenting costs?

This is not something the Patent Attorney wants you to think about. He tries to focus you on the premise that "If you don't protect it, you will never be able to claim it as yours" – and he is right. The difference is that protection doesn't need to involve a patent in the early development.

Not until you have independent validation of the market (how many people want to solve this problem, how much they are willing to pay, etc) and a proof-of-concept (evidence that your idea will solve the problem better and cheaper than anything else and there is a way to put this into these buyers hands for significantly less cost than they are prepared to pay). Otherwise you will have TOTAL OWNERSHIP of a worthless idea, and sadly, this is more than 98%

of patents filed in the USA since 1995. That's right, according to the US Patent Office, less than 2% of patents filed were able to recover their costs within the first 5 years of being granted.

So if patenting is not a priority, what can you do. The answer is you need immediate protection, but not immediate patenting – this can reduce the cost of protection to FREE.

Basic protection (before you have a commercial project) can be secured with non-disclosure agreements and/or partnership and collaboration agreements which stipulate ownership of any resulting IP. You may also need employee agreements which specify ownership of any ideas conceived or developed at work belong to the Company, and there are many others.

These agreements can become a very cumbersome process, just to customize, print and get new deeds signed and witnessed before you disclose. This process might be needed for up to 12 months and can involve hundreds of parties in a bigger project, but after the initial design of the deed, they are Free to implement. This doesn't mean you won't need a patent later, it's just that you can now push the patent costs out until immediately before you become commercial, so you have the provisional patent period to generate sufficient cash to pay for the full patent expenses.

The real trap for patenting is that once you start, you CANT STOP. If you patent lapses because you haven't paid your patent attorney, everything you have paid to date is wasted and worse, you have presented all details to the patent office for everyone to see and now you are not protected.

Another issue for patenting is that larger multi-nationals may scan the patent office for new applications and will simply challenge you for the sake of it. If your patent lapses (generally, you can't expect to fund a patent program AND litigation – no matter how frivolous – and they are betting on the fact that you can't afford litigation and patent expenses.)

Keeping your information to NDAs and Confidentiality Deeds will contain your secret longer, giving you more of a surprise to the market later. Yes, you run the risk of someone, somewhere coming up with something similar, but chances are, that was going to happen anyway and if you don't have $200,000 to fight it, your patent is undefended and therefore, worthless.

Likewise, getting a designer to make your solution look pretty is not going to help, if the pretty product can't make money. There are inventors who have been talked into building a fantastic prototype of a household product that in that form will cost $30 each to build, when the equivalent market price for these will be closer to $20 each at retail.

Every day of the year, there are toolmakers who are creating plastic injection molds for products that will never be financially viable. I have also seen a stockpile of some 20,000 units of a plastic product that cost more to make than the buyers expect to pay for it..... This was a $180,000 mistake that was on top of the $900,000 cost of design and tooling. It was always going to be hard to recover that project to profitability, given these early decisions.

So If we can say the design, tooling and inventories can be bundled together and placed in the "not until we have funding" category, then where should your priorities lie?

You should continue your R&D on a shoestring, because the value of your idea without market validation and proof-of-concept is practically zero anyway. Focus on completing your R&D with just savings or borrowing from your immediate friends and family, who might just consider this a donation, given the current project risk profile.

The first external investment should be in the preparation for a document to raise the funding for the commercialization phase. To do this, you need three things.

1. Independent Market Validation
2. Proof-of-concept
3. A transition audit

Without these three elements, your pitch is worthless. It doesn't matter how sexy your project is and how experienced you are at putting it together. Investors are not even going to look at the returns unless they know you have the risk under control. This is all proven in these three elements and without them, you are destined to pitch your sexy idea to hundreds of prospective investors over the next few years, until the funds you spending on patenting are not able to be maintained and your project slides into the public domain.

You need to understand these three critical elements intimately, BEFORE you start a project. This is the only way you can be assured you can meet the internal expenses all the way to pitch preparation. Once you have the process and costs under control, you can sequence more than one project at once and generate the big investments and returns every time. You need to know that spending $12,000 on market validation is going to save you millions in patenting, design, tooling and inventories - even if the research results say it's a dog.

No Compelling Pitch

From your five-second "elevator pitch" to your twelve-minute pitch deck, you have to be ready to present your case to people who (afterwards) really cannot afford to walk away. The elevator pitch was termed this based on the premise that you have the time it takes for an elevator to ride between floors, you can answer that "So, what do you do?" question. If you can summarise what you do in eight words or less, then you have an elevator pitch. If you can do that and have people ask for more information, then you have a good elevator pitch.

You need to have much more than a story, though. This is the business case for why you will build this and why you know people want to pay money for it. If your innovation is based around something you developed to satisfy a need of your own, pay attention to this section. Your business case should communicate why the public would buy your product. How will your product make or save your customers money? How will it save them time? Having a product that looks nicer than one that's already on the market may not be an advantage at all.

In the 1930s in the United States, there were thousands of new patents for mousetraps. Innovators assumed homeowners would want to purchase a better quality mousetrap. Many of these were elaborately designed and easy to use, but they lacked the one quality that most buyers wanted: disposability. Any market validation should have identified that no one wants to clean a used trap. Despite more elaborate design and better innovation, customers preferred a simple, cheap mousetrap that was disposable.

Recently, I worked with a company that developed an innovative method of coating french-fry chips to reduce the fat intake at cooking. The chips at most fast food chains are fried twice, once at the chip manufacturer before they are frozen and later at the fast food outlet before they are served. The project enlisted one major fast food outlet as a collaborative research partner, sharing the last mile of development for the tax concessions and a five-year exclusive first right of refusal to use the product in fast food outlets in one country.

After two years of development and three changes of leadership at the fast food company, the product was given a certified 35% reduction in fat. However, at that point the new CEO of the fast food outlet decided he did not want to be the chain promoting low-fat chips. His argument was that anyone who purchased fast food was not interested in low-fat products.

After careful consideration, the innovation company reconfigured their business model to focus on the savings in oil at the fast food outlet rather than the reduction of fat for the end buyer. Because the chips could be cooked faster on their second fry and they absorbed less fat in the deep fryer, the company was able to calculate the savings for each fast food outlet based on tonnes of chips cooked per year.

The net benefit for each of the outlets, particularly those that were franchised, was estimated between $12,000 and $28,000 per year. This became a compelling business case for the CEO to move the project forward. However distorted we as consumers may view this reasoning, the proposal to franchisees was an easier sell given that they were going to save money rather than have to persuade the public to consume low-fat french-fries.

The core issue here is that any business case must indicate for the investor what their advantage will be. At the cornerstone investment stage, few investors are only looking to increase savings. In most cases, they could and should be looking at investing in projects or technologies that will enhance their existing business or at least deliver a sustainable competitive advantage as part of their investment. What we think will be attractive to others may not be enough to convince them to invest.

I recently looked at a project that had developed a quick release fishing rod holder for beach fishing that could be adapted to rocks and boats. This was a clever mechanism to enable fishermen to have more than one rod extended into the water and to release quickly with one hand if a rod got a bite. The developer was a keen fisherman who developed a great product that he knew he would enjoy using, but he did not know if others would buy and use it.

When we did our market validation on the project, we identified that the only way the product would work commercially was to make it available to the mass market for under $20. Behaviourists suggest that because $20 is a single note, it takes less contemplation for people to dispose of it. Therefore when a product is under $20, it can be mass marketed with very little contemplation required by the buyer.

The target market for this product was the gift market and the serious fisherman, and the marketing channels we identified were fishing stores and the United States home shopping channel QVC, which can deliver tens of thousands of product sales in a day. The product is now in the final stages of development and will be offered in the spring on QVC.

Wrong Funding and Structure model

For most innovators, the main intention is to raise the funds to do the structuring. However, you need to get your structures right before you do the funding to ensure that you end up owning a piece of your project in the last mile of development. Elsewhere in this book I discussed the most common structures required for an innovation with national or international potential. If a project cannot do more than $1 million a year in its second year of operation, I generally would not participate. My modelling is based on having a one to $10,000,000 turnover in the second year of operation, which would be the first year of sales. If your project is outside these parameters, it may pay to ignore this.

The most sensible structure, even for a starter project, is a limited company structure. This is not a proprietary company in Australian terms, but one with public company accountabilities. This will give investors the confidence that you will be transparent in the operation of the company and with reporting at each interval. Three of the key reasons we do not have proprietary companies for commercialisation all have to do with accountability:

1. There should only be one director for the company, and that director need not hold any meetings at any time if he so chooses. If you cannot have a shareholders meeting, you cannot vote the directors out. This ensures the director remains in full control of the company at all times.

2. A proprietary company need not keep records or accounts. They are required to maintain records of transactions, but this may be bank account statements and nothing more. Consequently, shareholders will have access to no information unless they are signatories to the bank account.

3. The current proprietary limited company structure for shelf companies provides for the directors to choose not to transfer shares between shareholders. If the director chooses not to acknowledge that one shareholder sold out to another shareholder, the shares cannot be transferred and recorded as such. This can severely constrict a project if funds are low and a new investor is invited to join the project.

Too Many Concurrent Projects

Some innovators find a different or better way of doing something they have done throughout their careers. The majority of innovators, however, are creative people who can look laterally at problems and create innovative solutions that will save people time or money or make people time or money. In the majority of cases I have dealt with, innovators have a history of serial innovation that spans a range of fields and disciplines that are often unrelated.

This can present a major problem for investors. If an innovator is to raise early-stage venture capital or seed capital, investors may be less likely to give that innovator control of the project—including the budget—because they know that the money may be spent on other projects that interest them. From an investor's perspective, they have no interest in that other project and therefore do not want to fund it.

When things get tough for the innovator, he or she may look to other projects for small wins in order to gain motivation and momentum. This essentially means that the project being funded is parked for a short time and this other project becomes priority, often using the investor's funds. We refer to this as the "next shiny box" situation. It is easy when a project gets tough to switch to another project that looks easier. However, this may mean that neither project gets completed.

The key to avoiding problems is a contractual commitment for the innovator stating that no other project is to be worked on during the early-stage commercialisation of the research and development period being funded. I suggest including a clause stating that if the innovator does develop a separate project concurrently, then equivalent ownership in that project is automatically granted to the investor in the core project. This is usually an adequate incentive for the innovator to hold all other projects until milestones are met.

Most experienced IP investors will also look at where the project commercialisation will be housed. If the company plans to share office or factory space with other innovative projects with which the developer has a common interest, most astute investors would turn the project down. The reason for this is because they cannot guarantee that their time with the developer—which they are funding—is being used on their project.

Undefined Areas of Expertise

People inherently have comfort zones. We should be mindful of this, especially when funding projects to transition from one place to another. If a project has been in development for 12 to 18 months and has just received funding to transition to commercialisation, retaining the same people in this new phase of the project may not be justified.

If a project runs into difficulty during the commercialisation process, it is handy to have the development team available to rectify any issues. However, some issues end up being marketing problems that do not require further development to resolve. Given that most development teams are more comfortable with R&D, they may prefer to focus on further development of additional enhancements to resolve any issues that arise.

For example, in the case of the low-fat chips, the R&D team could have worked to create a chip with 50% less fat, but that may still not have sparked interest from the investor and licensee. The problem was essentially a marketing problem that took no further development to turn a good product into a brilliant product. The solution of highlighting the potential savings was universally appealing to all decision makers.

Complexities & Distractions

After years of development and investment cycles, many legacy projects have a convoluted legacy of shareholders, loans, promises, and other valid and invalid commitments. These must be addressed before the acceptance of any investment funds if the project is to have a chance of meeting its milestones.

The first issue to address is who claims what. All details must be tabled until there is no further claim to be made against any of the projects at a certain date. At this point, you should size the debt and equity claims to ensure a fair and equal value to all parties as the intellectual property is rolled into a new entity. This requires a unanimous agreement by all parties via circular resolution or a deed of company arrangement, if the project is a company or multiple companies. At this point, you generally would incorporate another legal entity and roll up all the intellectual property into this new entity, in exchange for equity and promises of future equity for the previous lenders, shareholders, and participants.

This may sound easy, but it can be tricky to implement. In most cases, astute shareholders will not accept the offer and may threaten to derail the project if they do not get what they are promised. Having an independent third-party negotiate this can be a significant advantage, as there is no room for an emotional appeal by shareholders to the principal driving the project.

In most cases, when the sizing is performed, there is a promise of future activity and reward to key personnel or others, for services or goods to be supplied in the future. These can be incorporated into the agreement, but they must not be issued as shares in advance. The best method of dealing with this is to have an agreement as part of the sizing, which promises shares or options at a later date for services performed.

Some of the most disciplined innovators can be distracted by personal and other business issues, which in turn will affect the productivity and development cycle of anything they want you to invest in. This is why most venture capitalists will ask about your family life when you pitch. They want to know that you are not fighting battles on the home front which will distract you from the development you are going to commit to using their money.

For most professional investors, key triggers would include imminent or recent divorce, pending or on-going custody issues, inter-department battles, obsessing with competitors, time-consuming hobbies (such as motor racing, etc) and even office romances. These type of issues can generate emotive responses in even the most rational of people and this can affect their physical and emotional availability, their decision-making and in some cases, their priorities. I have been told of a very bright innovator spent four days digging out his backyard, in preparation for a pending apocalypse. I have also seen another innovator refused to accept a funding check because the messenger had previously managed a company in which the innovator's father had invested and lost money. On the surface these can appear harmless and almost reasonable, but well may directly affect the development of a project because of the availability of the key innovator. Ultimately, everybody and everything in that project may be directly affected.

Not Enough Funding

One of the more common misconceptions about early-stage investment is that money is required upfront in a lump sum. Equity can and should be gained through the provision of cash and services as and when required by the project. Having the equity provider manage cash flow will generally give them some measure of control if a project is starting to get out of hand or the principals are failing to report progress regularly.

As part of my IP coaching practice, I provide inventors with help through a triage call process. This involves a 10 minute chat about what their current problem is, to see whether or not my experience can unstick where they are currently stuck. In two cases this week I was chatting with people with the same problem. Both of these had an excellent idea, but have invested all of their savings way too early, into protecting the intellectual property and building moulds and prototypes for full production. In both cases, they have not looked at the business model they needed to make this work.

Most inventors have limited funding and must choose very carefully where they spend their early funds. If their first call is to a design engineer, toolmaker, a patent attorney or other service providers, they will most likely spend their money on protecting or building their product, before they have even considered the commercial model it needs, to be able to bring this to market. Instead, I suggest they should hold onto their money and focus on firming up the demand that currently exists for the product.

It is a compelling argument that you will have nothing if you don't protect it, but equally, you don't need to patent an idea before you create the entire concept. Building your concept includes market validation, proof of concept, a protection plan and in most cases a prototype. Once you have all of this you can be assured - after testing it with the target market - that you will have a much better idea of how many people will want your product as the solution to their problem.

In costs practically nothing to validate your market and ensure that if you get this project going, you can actually make money out of it. I encourage inventors to hunt for problems and build a project that can fit as a solution. I encourage them to find out if the people who have the problem are prepared to spend money to "not have the problem". I also want to see what they're doing at the moment to solve the problem and how much that alternative solution is costing them.

Whether you have a prototype or concept ready to present or not, it's a great idea to bring a dozen of your prospective buyers into a room and conduct a focus group interview. This may not actually present anything you have developed yet and you might not even talk about your solution. Rather, you use that time to explore the problem they have and what their fears and frustrations are with having that problem.

If people have a problem and are prepared to live with it, it's not really a problem is just an inconvenience. In most cases, people won't pay money to get rid of an inconvenience and you can learn - well before you commit any serious money - that the project may not earn you the millions you expected.

With the cost of a focus group interview being the price of lunch for 10 to 12 people and the drafting of a nondisclosure agreement for each of them, you may have saved yourself $1 million in tooling, design and patent drafting, before you've actually worked out weather your Idea is a good one. This week alone, I have spoken with 2 inventors in this situation, both in excess of $500,000 and both facing more patent deadlines and trying desperately to sell or licence their patents with little or no positive sales track record.

Here is a question you should always ask yourself: "if I asked a tooling engineer, a design engineer or a patent attorney is my idea is worth pursuing - what is the only realistic answer they can afford to give me?"

Most patent attorneys I deal with are competent professionals and would not lead an enthusiastic inventor to spend their life savings on designing and lodging provisional patents, before any market validation is complete. However, with more new graduates coming into their industry, patent attorneys are becoming more and more pressured to nurture new prospective clients and the line between nurturing and pressuring can become blurred in some big firms.

The most critical factor in timing and sequencing is ensuring the project does not run out of money before either the next tranche of investment capital is available or when the project reaches a cash flow positive position in sales or licensing. No matter how generous an investor is at the early-stage, he or she will be less generous if the project runs out of funds before milestones are met. Expect that further capital may be provided at a substantial premium. This is because the principals do not have options and the investor does.

A project which has received funding but has subsequently run out of money is the worst type of project a prospective investor will see. Running out of funds is a clear signal that the principals are not competent in money management, project planning and/or coordinating resources. Any one of these three failures is a proposal killer.

If I were to invest in a project, I would need to be convinced that you know when the project will begin sales, when it will earn enough to repay me, and when it will earn enough to achieve the value realisation point that will give me the exit that I planned. Anything short of this is gambling, and there are few investors who will gamble on any unknown outcome.

This may be confusing to some innovators whose passion is to see the project survive. The investor, on the other hand, is looking for an independent authority to confirm that this product performs better than other solutions to the problem, at a far better price, in a far quicker time.

Presenting a cash flow forecast with unrealistic milestones in cash or timing or ultraconservative projections that yield a minimal return over an exceptionally long interval will not fly. The most suitable method of preparing cash projections is to have them independently prepared or validated by a recognised third-party. This may be an accounting firm, an auditor, or an industry specialist. The most important thing is to ensure that the data source you used to validate your cash flow claims has credibility with the investor. Having your cash flow forecast comparable with an independent market validation that you have commissioned will add credibility to your business case.

After the hard work of preparing a strategic business plan is done, many innovators still forget to provide a clear exit for investors. Although they invested because they trust you, investors still want to get a solid return within a certain timeframe so they can enjoy the benefits of their retained equity.

Sometimes just having an exit pathway will put your project above others being presented. Any sales or capital-raising document must be presented in the perspective of the investor if you want to get—and keep—the investor's attention. Many investors get comfortable with a project that is progressing to milestones and decide to stay beyond the first exit point. Some even pour in more funds as a project progresses. The key to success here is to point out all potential exit points so the investor has options.

Sometimes you only get one chance to do something right. With the commercialisation process for intellectual property, this is more the rule than the exception. Being forewarned about the seven deal-killers and making preparations to avoid them will give your project a significant advantage over other projects for equity and debt funding, access to commercialisation grants and other collaborative opportunities that will make a difference in your project's ultimate success.

Understanding what you need before you need it will help you plan the appropriate paperwork, strategy, negotiation talking points and collaborative research partnerships. Getting completely comfortable with these will set you up for future success in early-stage commercialisation.

Becoming the hunter

One of the common themes across many of the networking groups and forums that I see, is the expectation of many inventors that *"...if I tell my story to enough people, someone will want to get involved"*. For most of these projects, raising the required capital or engagement of a professional to raise this for them, just doesn't happen.

So.... given the premise that if we repeat an action and expect a different result, we are pushing towards insanity – I would like to provide some brutal clarity for you. If you have a really advanced, ground-breaking new idea that just needs one million (insert denomination here) to get going, your chances of succeeding are less to do with the project and more to do with funding. For most inventors, funding is almost impossible, because they can't get investors to say yes.

But, what if you learned how to change the way you look for funding, in order to get the outcome you need?

I propose that many of the less successful inventors (with equally great innovations) are disadvantaging themselves because they don't know how to hunt. There are 4 major types of funding approaches and in most cases, only one will bring you the result you need. If you pick the wrong approach, you will rarely attract what you need and this is made worse when you decide you should compromise the offer or concede far more of the project, in order to attract even more people to look at the offer. It rarely works.

The first of these four categories is the **Shopkeeper**. I will not spend much time on this, as this is usually the domain of larger corporates, who believe in presenting their IP opportunities as a "Business for Sale" and hope that third parties (brokers) may attract interested parties to partner up with them. This is low-cost and very passive, but it only works when the offeror is a successful company and in most cases, the investor chooses to become involved because of the partner, not the IP.....

Next we have the **Farmer**, who starts down the self-funded commercialisation road independently, hoping that investors will turn up as he/she gets more traction. This can be effective......if you have years. Some projects may take a decade to grow-out and in the meantime, you are carrying all the risk and possibly having to work for someone else to pay the mortgage. If you have patents, each year you delay your acceleration is reducing your value, so this method becomes less attractive given the time value of money.

The most common category is the **Beggar**. Before you dismiss this as not applicable to you, just think of how you present what you have, to new or uninterested parties. Are you telling everyone you meet that you need money? Do you tell them how good your project is but all you need is that extra little million to fast-track it? Are you asking accountants, lawyers, consultants and/or patent attorneys to provide you with services or their time, for a small piece of what they know nothing about? Worse still are you asking for a chunk of their time for a success-fee?

How different is this to begging? How many people do you know that set out every morning of their lives to look for beggars to hear their story and give them something for food? If you talk to successful inventors, they will all tell you they became successful when they stopped begging and honed their hunting skills.

So, let's look at the **Hunter**. If you stepped into the wilds with a spear and a hunger, you will still starve to death. You may have some exciting times chasing game and picking fruits and berries that may or may not kill you. But.... If you decide to acquire some hunting skills and then build a plan, you might just become the Apex predator in your corner of the wilderness.

So, if you accept that - right now - you do not have the hunting skills you need to bag that elusive investor, then your choice is to either get the skills or engage with someone that has the skills.

So what is different with inventions? If you stop and think about it, could you really expect any professional service provider to give you his/her time without payment? Why should they take on your project risk for a minority shareholding of an unproven concept, with someone they don't know? If they actually said yes, you can guarantee they are desperate and starving and possibly NOT the right people you need.

If you have read this article to here, you may be starting to understand that you need to become a skilled hunter or at least have the cash to engage one. In truth, you are facing just three options from here:

1. You have the funds to pay a professional, who will get you what you need under a "Done-For-You" business model. This will never be commission-only or "success-fee" based engagement. Even the Brokers who sell equity at listing for public companies, charge several hundred thousand up-front.

2. Learn how to be a hunter, by spending the next year or two studying the hunting game, the prey, the successful hunters, and then acquire the weapons that will make it possible, or

3. Invest in some structured learning that can reduce those years to months, or even weeks.

It still means that under Options 2 and 3 you are ultimately going to do the hunting, because you can't afford to pay a professional hunter. However, the skills are going to last you well beyond your first successful hunting expedition. The only question now is are you OK to spend the next year or two learning how (in my case) the capital-raising game is played, or are you prepared to bite the bullet and pay for some professional training?

I receive about 40 solicitations for services a month, from passionate inventors who can't seem to crack that funding code and expect me to get involved for a success fee. All this happens because they don't want to bother to learn about how to hunt properly.

So, if you have a red-hot innovation and you can't get funding without some expert help from a consultant, an accountant, a patent attorney or a lawyer, please consider their likely response if you ask them to provide their limited time and services to a high-risk (currently unprofitable) project, for a minority equity or success-fee. It doesn't happen.

CHAPTER 4

The Funding Search

"Chase the vision, not the money; the money will end up following you."

—**Tony Hsieh**

Every time I see another article about a "Startup" raising several million dollars I internally brace myself for the inevitable phone calls from inventors and R&D team leaders who want these same savvy investors to throw that sort of money at their new idea.

The reason why it doesn't happen comes down to a poor choice of words in the Innovation industry in the USA. I am sure a little understanding will change all of that for smaller innovators who are not at the scaling-up stage of their business operation. Let's try to put some perspective to the differences of terms used by developers and venture capitalists.

When you read about these Start-Up firms raising millions of dollars for starting up, it's actually for advanced business ventures looking to SCALE-UP. This is a big difference. In every case, they have surpassed the requirements of a true Start-up and are at the tail-end of the commercialisation pathway - ready to launch globally.

A venture capital firm views start-ups as a business with a solid, positive trading record, with a substantial potential to scale the business geographically and/or demographically. They will generally look at businesses where the model can deliver a 100-fold increase in sales and profits, given the right funding and other resources they can contribute.

For early-stage commercialisation, we refer to Start-ups as projects which have completed the ideation phase and are about to start to firm up the concept and build the business model which will commercialize the intellectual property, once it is owned. For the Innovators who are just starting out and would classify themselves as Start-Ups, let me explain the stages without any complexity. First, the model (there is always a model).

Stages of Development							Capital Sources
						7. Live it up	IPO or Trade Sale
					6. Sell it up		
				5. Scale it up			Venture Capital
			4. Grow it Up				
		3. Stitch it Up					Cornerstone Investor
	2. Prove it Up						
1. Start it Up							Friends & Family

Fig 1: stages of Early Stage Commercialisation explained

1. Start it Up

This is traditionally where you have an idea and you want to firm up the value model and examine where this would fit (market demand, competition, pricing, costing, etc). You might have some sketches of the object on a napkin or perhaps you have a flow-chart of how it works for the people who need it. You have spent no money so far and the only essential issue is non-disclosure, for any party you share this with. You have not accumulated any tangible value so far, but your cursory investigations suggest people will pay money for this and you should be able to make it for a fraction of what they would be willing to pay.

2. Prove it up

Now we want to build some certainty into the concept. We search for published information on the market (buyers with the problem you solve), the industry (competitors who solve this in other ways), similar claims or patents (held by others) and the potential, how much people spend to solve the problem – locally, nationally and globally.

The most critical feature of this stage is the business model. You have to have a very clear process of creating, storing, selling, delivering, installing, servicing and supporting whatever this concept is.

This is also the point where a prototype might be constructed, to prove the concept works. Outcomes would be (1) an independent market validation, (2) an independent proof of concept and (3) a transition audit – where a qualified, independent party can verify that you have everything you require to build, sell and support your concept, given the appropriate amount of funding.

3. Stitch it Up

At this point we will need to get some ownership into the project. This usually starts with agreements with all parties you will seek help from, to ensure there are no external claims to ownership at a later date. This may be complemented by copyrights, applications for provisional patents, cooperation and/or collaboration agreements as well as staff and supplier non-disclosure agreements. You should now have an asset, which as yet, may not be worth much, but is owned. Most projects would now incorporate a separate corporate entity to own the intellectual property, as this will be easier to manage value and risk at a later date.

Once you reach the end of this stage, you are ready to search for a Cornerstone Investor. This is traditionally a single person or entity who is heavily vested in the market or industry and who can obtain a competitive advantage for their current operations, if they are able to become a minor shareholder and technical (or marketing) contributor to the project.

4. Grow it Up

This is they stage where our project begins to trade. Our cornerstone investment is designed to get us from here to the scale-up ready status, so we have to prove our product and the business model both work, as well as both being able to be scaled to meet the (now qualified) national and global market demand.

Think of this as the first test drive after you have built the race car. You want to run it through a few laps and make sure it will be competitive when you ask the big sponsors for commitment. You should be able to source your supplies, create product in production runs, establish and feed distribution channels, and have sufficient profits to prove that this project will make money, even at its minimal production levels.

Traditionally, venture capitalists will be looking at this phase of the operation to determine if you can obtain a significant market penetration with useful margins, in a local market. You are better off having a larger share of the local market than a minuscule portion of a global market, even if the latter is more profitable in the short run. At this point, you should have a profitable business operating in a single market, with an significant and obvious potential for scale. This is what will attract the venture capitalists and/or private equity firms, who want to add their funding, skills and contacts to ratchet up your productivity by 100 fold. They would classify you as Start-Up ready.....

5. Scale it Up

At the commencement of this stage, we are where the venture capitalists refer to as start-up. In fact, you have an operating business which is profitable and yielding a steady but consistent growth. There are some innovators who consider that they can grow their business organically, once they reach this stage. The attraction of venture capital is that you can achieve 10 years growth in one year, and global markets in the time it will take you to get national markets, with a tenfold to 100 fold increase production, which would yield you significant economies of scale. All of these add substantially to the bottom line and any good innovator should be prepared to share this bottom-line with a venture capital firm, if they can deliver those sort of attributes from their partnership.

6. Sell it Up

Venture capital firms understand that a project has a life span. For these funding and growth firms, they are only interested in the high-growth phase of the operation and would plan their exit well before they begin. This may mean that they would like to have your project sold at a particular point, so they can realise their investment and take the profits to their next project. They do not want to be married to your project for life and will set a window of perhaps 1 to 3 years, before they would seek to dispose of their equity position.

Their exit may be in many different forms. These will include a management buyout, where you leverage the assets of the company to borrow sufficient funds to buy their equity yourself. They may even give you a vendor finance package, with security over the firm. Other exit options may include an agreed trade sale to a third party, which may include the entire operation as a strategic business unit for the buyer. This would require the services of your team to remain within the project. The upside to this would be that you could receive shares in a public listed company, which can be traded as cash.

Another popular option is to list the company on the stock exchange in the local jurisdiction. There are requirements for each different stock exchange in terms of size, number of shareholders, turnover and profit, in order for the project to qualify for a listing application. Most stock exchange applications will require a rigorous due diligence process, which is very expensive and can take 6 to 12 months to fulfil. As exits go, the venture capital firms can then choose to liquidate their stock at different points in the market cycle, given that the stock is then publicly traded.

7. Live it up

Once your project is fully funded and you have your exit implemented, you may still own the project or a substantial part of it. As a mature trading company, your role may change to a senior executive in a large public firm and this may not fit your lifestyle. You may choose to retire from the company and play an active role in further developments or take on an external role to acquire similar businesses in which you can add value in research and development to create new income streams for your now public IP company.

So where does the confusion lie? As an innovator, you would see start-ups as the ideation stage which is generally in research and prior to development. The capital markets see start-ups as operating businesses with sensible profit margins who could scale the business model geographically and demographically to acquire a tenfold to 100 fold increase in their current operational capacity. This process can become very frustrating when innovators spend days or weeks preparing a business plan for venture capitalists, when the project is still at the ideation phase.

If you are looking to fund an idea, and somebody tells you that venture capitalists, government or investors are looking for start-up projects to fund, you will understand that your project is not ready for scaling up. These funding entities are actually looking for profitable, operating businesses with intellectual property that has a high degree of innovation and sustainable competitive advantage, with the business model that lends itself to scaling. Your capital requirements at ideation are more suited to donations from friends and family, until you have a proven concept, with independent market validation and you have conducted a transition audit to determine that you have all elements of your project available for inspection. At that point, you're still only at the cornerstone investor stage, but you will need this type of capital to set the growth trajectory and to capture markets as you put your project into production.

What do you need to pitch?

In the past few years, I have seen several projects which have prepared very sophisticated offer documents for funding, but have not been funded because they overlooked the most fundamental of tenants – to play to their target market.

In most cases, when investors are looking for early-stage funding for ideation, they don't seek outside help. This makes sense, when you consider that the risk is far too high for external investors and the amount of money required is usually fairly low, given that it is rarely recommended that investment in patenting, tooling or production is considered at this time.

The funding point at which most inventors become stuck, is looking for the cornerstone investor. This is the point where the innovation has completed research and development and is ready to transition into the early-stage commercialisation phase. The assumption that can be made is that because the product definition and market validation is complete, there is a tangible value on the asset and as a consequence, and so a provisional payment has been applied for.

The cornerstone investment funding is usually around $500,000 to $1.5 million and is best suited to a single entity which is committed to the target industry or market. This person is usually a successful player in the same industry or a retailer/distributor of similar or complementary products. In almost every circumstance I had anticipated in over the past 20 years, I have not seen a cornerstone investor who represents a highly sophisticated or professional investor model. These people are passionate about their market and industry and will get passionate about your product if it can mean more competitive advantage to what they currently offer in the field in which they play.

Given that most cornerstone investors are not sophisticated or professional investors, my experience suggests that most of them would look at a sophisticated offer document as a bit too slick for them. They have to feel comfortable with partners they get involved with and if you present a double-A round funding document based on a US high-growth company (when you are looking for funding within Australia for a small commercialisation project) they will generally tend to pass on the project based on the perception that you may know a little bit more than they do about corporate funding.

The average cornerstone investor has expertise in the commercialisation, marketing, distribution or production of the type of products that you are developing. They rarely have experience in international stock exchanges or securities platforms. So why would we present an extremely sophisticated document to a relatively unsophisticated investor? It doesn't make sense, right? Given that our objective is to get a yes to what we want, then we should present what we want, in a format that they will not feel uneasy about.

So as much as your glossy prospectus-style offer document impresses you and the accountant to prepared it, it is likely to have the opposite effect on the person it has been designed to impress. It is no accident that many projects get designed and funded on the back of a napkin, in the favourite restaurant of the investor or partner. Short, sharp and sweet is the theme and the foremost thought in preparation of any document for a cornerstone investor is to only present the information required and put everything else aside for the due diligence files later.

After some 15 years of presentations to single investors and corporate investors, I learned the difference between impressive documents and successful documents. I prefer the latter. I use my experience to build a 12 point presentation to cornerstone investor prospects, which can and should be presented in less than 20 minutes. The key is that I only present the information they require to make the decision and I provide everything else upon request, in the due diligence file. The DD file is subject to some deed of nondisclosure, although the formal presentation rarely is.

The secret to preparing the right presentation, and pitching to the right person at the right time, is an art in itself. If you have an R&D project entering the commercialisation process and you haven't yet locked in your funding, then you need to get into the mind of experienced investors and see the key principles from their perspective.

In my 28 years of commercialising intellectual property, I have been presented with hundreds of projects which stumbled at this roadblock, perceived by many as the first. The first roadblock is actually not getting debt or equity finance to complete the development phase and move on to commercialisation.

Most innovators who fail to secure investors early, believe the difficulties they have are around risks and returns, but the main reasons for failure are far more complex. In this chapter, I want to outline some of these issues and present what I call must-haves for getting funding into early-stage commercialisation projects. What you are trying to do is meet the many requirements of sophisticated or professional investors. If you meet those requirements, then they will feel encouraged to invest in you and your projects.

Let's begin with my top five must-haves to produce immediate and positive change in the way your project proposals will be received.

1. Structure

The first key element is the right corporate structure. R&D projects need to have the appropriate accountabilities by the managers, so the right corporate structure makes the directors accountable to the shareholders, in most jurisdictions. In Australia, we recommend a public, unlisted entity, which requires the following:

- Three directors (with at least one independent)
- Formal audited accounts
- An independently maintained share register

Under Australian law, a director of a proprietary company can be a sole director and can also occupy the company secretarial role. This essentially makes him or her unaccountable to everyone else, including shareholders. The only requirement for records is to maintain income and expenditures for tax purposes. In this case a bank statement will suffice.

Directors have no obligation to provide a copy of any financials to shareholders, but they can choose to do so. In some cases, shares have failed to be transferred because the director was under no obligation by law to transfer shares that had been transacted. Any astute investor will be aware of these conditions and will seek a safer structure, with accountability and transparency, for his or her investment.

More than 20% of the projects presented to me for assessment are projects which have already received some funding and for whatever reason require additional funding. Sadly, if the structure wasn't appropriate before the first investment was received, it is almost impossible to cater for future investment unless 100% of the investors agree to the terms of the change. Getting 100% agreement can be difficult.

Recognise, however, that coming back for additional funding through structural change alters the strategic dynamics among seasoned investors. In larger projects, some investors will try to leverage this new information to better themselves against the others, ultimately imploding the project.

I was recently involved in a complicated project that shows the difficulty of restructuring. The project crossed three different countries with three different corporate structures in three different denominations. The project had raised money at different prices from different investors. Even keeping track of these iterations was complicated! But before we could get traction with further investment, we had to restructure the whole entity with a single entity with a single domicile so a public listing could become a reality.

This work took nearly ten months as minor parties jostled for advantage and some parties obstructed just for leverage. This extended the burn rate for the commercialisation process, and although the project had its first eight-figure sale, the rollout was approximately a year longer than planned.

Lesson to be learned: Keep structure at the forefront of your plans or face the consequences later.

2. A Planned Exit

At the bright and positive start of your project, this next point may sound negative or foreboding. Experienced investors looking to jump into your project, will also want to know when they can jump out. The orchestrated exit must be a feature of every investment proposal.

Without a clearly defined exit, advisors will tell their clients (your prospective investors) to eye liquidity and buy public stocks, which can be traded at any time. There's no standard exit timeframe, but depending on the lifecycle of the project or the size of the reward, the project can offer a six to 24 months exit to a Cornerstone Investor. At the exit point, one or all parties need not necessarily dispose of their existing equity. They simply must have clear options presented.

A less controversial term for exit may be 'value realisation point.' This is a defined interval in the development process or commercialisation process, where values can be determined absolutely and parties can buy or sell shares or other securities within the project. This does not necessarily have to be a public listing.

I dealt with this exit condition in a project that aggregated a group of property rent rolls, spread across Australia, bringing them together for a public listing. The commercial intent for each individual owner was to provide a public listing on the Australian stock exchange as an exit point. Where no individual property manager company was large enough to list in its own right, the aggregation of a number in each city, made perfect sense.

We were able to raise the seed capital to put the financial structure and corporate structure together, which then enabled us to bring all parties to a public listing under a suitable document. This meant the principal as well as the individual owners obtained a higher-multiple valuation as a public company than as individual private companies.

3. The Right Team

In any project, the structure and experience of the team is going to be a key component of success. If you assume everything that can go wrong will go wrong, then your team will be the difference. In most cases, experienced IP investors look at five elements of a project to determine its attractiveness:

1. The People
2. The Product
3. The Industry
4. The Markets
5. The People

That's right - the viability begins and ends with the people. Most experienced IP investors know they can throw other opportunities at the right team, to turn a dog into a fox. On this point, you need to demonstrate relevant experience in the roles of R&D and commercialisation. If you haven't separated the roles, the investors will wonder why they should fund your learning curve, or worse, they will wonder if you will be able to maintain the same rate of growth that you have achieved in the development phase.

For the sake of practicality, you need to have separate teams with separate skill sets, accountabilities, and most importantly, separate budgets. Show the investor how each position is to be filled by the right person for that role, not simply filled according to the person's knowledge of the product or history with the R&D program.

I have witnessed many cases of teams receiving funding before my engagement and running out of money because the project team had come up against commercialisation roadblocks. Their response was to push the project back into R&D to build an iteration to overcome the roadblocks. This will eventually cycle the project back into the comfort zone of the R&D team which, of course, is the R&D process!

I recently reviewed a project with an extremely good product that received a lot of interest from government buyers – it was even already receiving sales. The project needed capital to step up the commercialisation process. However, during the scoping interview, the principal did not want to relinquish control of the project or the company that was to own and drive the project. He also wanted to have this project reflect his achievement, so he could be recognised by his peers.

After careful questioning, I recognised that his intention was to drive the commercialisation process from within his research and development framework. I saw that the project would not achieve its intended goals unless it had the right people to drive it. I decided not to take on this project, which is disappointing because the product was an excellent and unique solution to a readily-identifiable problem and by not including this project I wasn't able to rescue some friends who had already invested in it.

4. The Right Amount of Funding

The key with funding is to not ask for too much, but never ask for too little. One way to temper your funding request is to identify costs that can be pushed out. These costs might be paid out of future cash flow, as the value of money at an early stage of the project is far higher than later. The project team should only seek funding (for either debt or equity) for essential items critical to bringing the project to a positive cash flow.

One of the biggest issues in commercialisation is always budgeting. If a project is under-budgeted, the money will generally run out before the project is commercial and everybody's investment will be at risk. If a project is over-budgeted, the foundation shareholders are losing.

If, after the expenditure of these funds your project is still incomplete, then you will need to be extremely nice or ruthlessly tactical, to dig your project out from under the ensuing chaos. Worse still, some investors may fund you with the intention of running you out of money, so they can offer the additional finance at a much steeper rate, thereby taking over your project.

A major stumbling block will come from weak planning documentation. If you have not prepared your budgets and researched your burn rates, or you haven't timed your patenting expenses and so on, most professional investors will walk away from the project.

In most cases, it's better to present your projections in a statement of source and allocation of funds. This allows you to put your cash flow projections alongside your equity placement or debt placement as agreed. Investors can assess at one glance how the cash flow will track over the life of the commercialisation process.

About a year ago, I looked at a promising project by a group of software developers, a project I believed was truly a global game-changer. I did some rough costings so I could get an idea of how much they actually needed to drive the project to a commercial reality.

They did not engage me at that time however, since one of the prospective investors objected to spending money on an independent consultant. The project principals made an agreement with a Cornerstone Investor for a commitment that was substantially less than what was required, under the investment calculations we had made. The project subsequently ran out of money and everybody was stuck.

The foundation shareholders had more than they could afford to lose tied up in the project, and the Cornerstone Investor had a considerable amount of their investment buried in there too. It wasn't until all parties were able to negotiate an internal agreement on corporate valuation that the project was able to prepare an offer document to raise capital to reach a cash flow positive position. In VC (Venture Capital) jargon, everybody had to "take a haircut," which essentially meant revaluing each of the shares issued to accommodate the equity issued to new investors.

5. Reward for Money and Time

Naturally, a Cornerstone Investor in your project wants a reward at the end of a defined period. In most cases, project leaders present their R&D project for commercialisation with a mind to the returns. However, rewards change in value depending on time – it is essential then to remember that money has a time value. If your returns for a Cornerstone Investor (this represents the highest risk of early-stage commercialisation) are about a 100% increase on equity placed, then there will be two separate values if the returns will be received in six months, or two years.

In most of our commercialisation projects, we turn to a Cornerstone Investor who has a commitment to the industry and/or the markets, but we would only ever place them into a commercialisation project for a period of six months or 18 months, depending on the project and the returns.

Why these two periods? Well, most successful Cornerstone Investors are also looking for leverage, for their own projects or for market advantage. They may be a buyer, user, or component supplier to your project, and they will stand to benefit from additional sales for every unit you produce and distribute. They will want to limit their investment period but extend their engagement as shareholders. In some cases, it is prudent to offer a measure of exclusivity for a limited period, as a sweetener for the investor.

Without absolute certainty in the calculation and assessment of rewards and risks, it is very difficult for a commercialisation team to present any project. In the prospectus or other public capital document, independent parties do this due diligence to assure investors the evidence of any claim.

Many project principals do not calculate the lead-times in getting a project to market. In contrast, smart project management teams plan out all resources, including the timing of those resources, for costing and lead-times.

An astute investor will look for how the project is presented logistically. This will show how thorough the developers have considered the allocation of resources and the natural bottlenecks that occur in any project.

I recall an aggregation project several years ago in which we spent months pulling 12 parties together for a public listing. After the first phase, the principal felt ambitious and included another large entity into the mix before the listing. The due diligence alone on this task created an additional three-month lag for all the original principals who had become foundation shareholders.

One of those foundation shareholders approached me because his business was now tied up in the pre-float due diligence process while his wife was battling cancer. He could not retire from the business to look after her. I was powerless to intervene, as the managing director of the merged project took it upon himself to extend his offer to several other parties. He wanted to make a bigger impact on the financial markets at listing.

This behaviour was called to account by the capital markets when the IPO was announced. The target listing price was not achieved nor was the target share price after listing due to the delays reducing value and wasting money. Yes, the managing director found additional funds, but the delays took the value back to square one. Let's now turn to a secondary set of observations that are also important.

Funding Increments

At first you may want all the funding up front to provide security. You do not have to accept 100% of the cash on day one. In fact, receiving incremental payments based on milestones will give you a much better cash management system, reduce your risk for development staff going crazy on unbudgeted subprojects, and offer the investor some measure of accountability.

One Verses Many

In early-stage commercialisation, I believe it's better to have one investing entity in the mix. Things do not always go perfectly, and having one investor to liaise with saves a lot of commercialisation resources and time for investor relations.

With just the one investor, you can more easily change direction if something you trial does not work as planned. You may have targeted a particular market segment that is proving difficult to get into. If that's the case, then sometimes it's better to just cut your losses and take on a different demographic. However, if you have many shareholders, it's hard to find consensus on change of direction.

This is not a hard and fast rule – sometimes it's helpful to have more than one. The advantage of having many investors is that no one party has excessive decision-making influence during the early stages of a project. This could be a tremendous advantage if you are developing a project in stages and would like to raise additional capital for the next phase. If you were to deal with a single investor, then it would be much more difficult to raise more capital if you ran out.

These two options may be combined. My view is that you should have a single Cornerstone Investor for the research and development phase, and the same or another Cornerstone Investor for the early-stage commercialisation process. When you step up to a company, you should seek to spread the shareholding to over twenty0 people so you can take public funds at a later stage.

At the conclusion of the commercialisation process, your project is worth a considerable amount of money and you do want to maintain momentum without having control battles. With this in mind, it would be easier for a project transition into commercialisation by focusing on issuing a wider document for capital-raising, which can attract many investors for a manageable proportion each. At that point, you need to include investor relations in your budget. You may handle this job from your website, but the more information you put in and the more up-to-date that information is, the less concern your shareholders will have going forward.

Timing of the Draw-Down

If you have commitments for funding the commercialisation process, then you should examine whether or not you need all of this money upfront. It could be that you receive a commitment for a 12-month drawdown against agreed milestones. It will always be prudent to have a contingency budget set aside that the technical advisory panel can allocate as they see fit.

The technical advisory panel should determine whether the milestones have been met, and therefore allocate the transfer of funds into the commercialisation account. If the Cornerstone Investor has not provided the cash up front, then there must be penalties assessed on the investor for not providing the funds on time. Delays in investment funding can be devastating for a commercialisation project. Several years ago I was involved in a project that raised $500,000 for the completion of a developmental program from a single entity, consisting of two businessmen. Subsequently, they had a falling out which allowed one to take over the project.

This gentleman was involved in property development and lost a lot of money on a project immediately after he invested his first round of capital. He had placed $100,000 into our project, with another $400,000 scheduled over a 12-month period.

Given that the following $400,000 seemed a certainty, the project leaders approved a 12-month lease on commercial factory to build a prototype with security. When the sole investor could not meet the requirements for the further $400,000, the project had to be reassessed. As equity had only been allocated on a supply basis, no shares had to be recalled. However, the project had leased premises for twelve months but could not afford to do any R&D, over and above the $60,000 that they put towards components and test equipment.

Two years have passed since that project went off the rails. Significant restructuring has been done to get the project back on track. In the meantime, some $40,000 to $50,000 of that initial investment – to pay for a factory that could not be used – was wasted.

Financial Offsets

Although in most cases cash is always going to be the best commodity for equity, there are times when one can trade other items of equivalent value to speed up the development process or the commercialisation process. In the past 28-plus years, I have managed to negotiate access to plants and equipment, trading the available tax concessions that the company in development could not use because they were not in a profit situation.

My most creative project for equity trade was to provide equity in a project for the business associated with that number one client. The client provided several million dollars' worth of services under a long-term contract in order to receive 30% equity and then purchased another 10% equity at a given price.

Another creative way is to form a collaborative research partnership, which becomes responsible for the expenditure on development and/or commercialisation functions. If one or more of the parties pays the accounts as and when they become due, they become development expenses or commercialisation expenses and therefore could provide a company with a tax deduction. In some cases, the concessions provided under commercialisation or research and development can be much higher or matched dollar for dollar by some governments.

The Cornerstone Investor Model

Throughout this chapter I've hinted at the importance of the Cornerstone Investor. Now I want to provide more details about the ideal type of Cornerstone Investor for the commercialisation process of a project. The ideal Cornerstone Investor is as much attracted to the market or industry as to the return on investment. The Cornerstone Investor should be:

1. Committed to the industry and/or market that will give that entity a competitive advantage through exclusivity or vertical integration.

2. A single entity or person, because one of the key ingredients will be passion and this is hard to engender in a corporation.

3. Capable of making their own decision on the investment. If for any reason they outsource the due diligence, it is rare for any third party to approve the investment because they will have vested interest elsewhere and the only outcome for them is to wear the risk of a positive decision.

4. A sophisticated investor capable of providing at least $500,000. If the project cannot be valued at more than $1 million at that point, then the cornerstone investment model probably is not the appropriate model for the project. (Note, if you are seeking a small amount, such as $50,000, then you should get this from your own bank manager or friends and family. The returns on this sort of investment are not going to attract any commercial entity or the sophisticated investor, because the amount of time they need to allocate to the due diligence will more than eat the profits the project could possibly return.)

Understanding other Transaction Currencies

When dealing with an investment proposal for commercialisation, it really pays to understand the difference in transaction currencies associated with all the parties involved. For instance, my time with Universities taught me that in most cases parties in Universities are not in the slightest bit interested in profit. They do not get to keep the profit and, in some cases, do not even get to spend. They are more interested in recognition and they judge this recognition on lab assistants, equipment, board positions and grant funding. Therefore, you can get more done with one or two lab assistants for a year (paid salaries) than you could with $1 million cash. Understanding this difference can ensure your project gets over the line or doesn't.

I also learned early on that academics do not compete with the outside world or commercial entities. An academic researcher competes mostly against the person sitting in the cubicle or office next door. That person applies to the same department, for a share of the same budget, and if his or her projects are more meritorious, then they will be funded over yours. When a project approaches a University academic for some collaborative research and/or access to some specialist equipment, being mindful of the desired outcomes of the academic will significantly help you in negotiations.

I once was appointed to conduct an audit of a State Government (all departments) in relation to intellectual property (IP). In the conduct of this IP audit, we identified that a health department (which operates hospitals) was not providing any intellectual property for commercialisation – an unusual situation. What was going on? It turns out Universities had approached these hospitals and offered the heads of the departments adjunct professorships if they assigned the research to the University. As that department had no methodology of capturing and then rewarding these heads of department for research conducted in the department, the IP leakage was rampant. We estimated that the losses were in excess of $25 million over nine years. This occurred because the University commercialisation department recognised the currency which Department Heads were willing to negotiate in, was not cash.

In just about every deal you contemplate, your work would be much easier if you accept that most transactions do not have to be all cash. If you trade your negotiating points for the services or assets you intend to buy with that cash, you will reduce the cash component of every deal and make it so much easier to complete. It may come as a surprise for many inventors, but many of the commercialization deals being negotiated today are completed as cashless transactions.

There are several transaction points In the IP commercialisation journey, in which third parties will provide some consideration to be part of your project or to licence your intellectual property. Arguably, negotiating for cash to commercialize is one of the hardest elements of the IP commercialization process.

It doesn't have to be this hard. Half of the struggle is the qualification of the potential investors. Most inventors will struggle to negotiate a trade sale of equity in their new project, for a bunch of money to buy or rent resources to get the commercialization done. What we generally tend to overlook is that the people with those resources would be more willing to give us access to those resources than part with money.

Acquiring IP without cash

Likewise, this can happen in the IP acquisition process, where you might be seeking access to a project that has been developed within a Government or tertiary environment and there may be no actual mechanism to accept money for the IP. Several years ago, I conducted an IP opportunity audit of a State Government in Australia and in that process we identified one teaching hospital which was part of their Health Department. They had their own innovation developed in their course of business, but the Health Department had no formal process to commercialize it. This opened up an opportunity for a University commercialisation team to offer these inventors an Adjunct Professorship if they assigned the IP to the University. In some cases they also funded equipment required to complete the development, which then gave them an endless feed of almost-completed projects - for next to nothing.

It could be that you identify some innovation that would complement yours and you want to negotiate with a third party to secure rights in your segment of the market. The most fundamental (but often overlooked) fact is that cash is not the only currency. In fact, in the deals I have done over the last 6 months, there has not been one deal that was 100% cash as consideration.

One key factor in this type of transaction is to have done your research before you get into the meeting.

Situations

You may be involved in buying or licensing IP from Universities or other research-focused organisations. You may be involved in seeking funding and/or other resources for your development and/or commercialisation and you make be negotiating a licence, a trade-sale or a public listing to fund a project growth program.

I recently worked with an advanced wound care project looking for something unique to add to their portfolio. They had a single product and were looking to become a serious player in this industry, where the top 5 players in the USA were all posting annual sales in excess of $10b per year. I found them a small company making collagen from animal skins, in a disease-free country. For the collagen company to expand into medical-grade collagen they needed secure markets and for this wound-care company to crack that market, they needed assured, exclusive supply. No money was exchanged. Instead the buyer developed a clinical protocol for extracting all of the endotoxins from the raw product and granted this process to the supplier, in exchange for exclusivity. Once volumes were established, the investment into this collaborative venture by each individual player was assured. This type of venture requires no money (assured markets can make debt finance very easy) and nobody had to dilute their ownership in their respective operations.

Types of Currencies

When firms are about to negotiate IP with a University - in most cases - the price the University commercial department have placed on the project IP could be a pure guess. I can therefore prepare a counter-offer which rewards the Department Head, who is generally the key decision-maker. Imagine securing some promising science innovation for the price of some lab equipment and 2-3 research assistant salaries per year. By rewarding the Department directly, the value becomes far more to the Department than any cash transaction to the University. Some of the more creative currencies I have deal with include:

Use of Facilities: Secure lab facilities and use of equipment — with or without the skilled operators.

Professional/Peer Recognition: I have mentioned the example of granting Adjunct Professorships previously.

Sponsorship of a paper to be presented at a World Conference: We all have egos and in an industry based on recognition as the currency for promotion, being sponsored to prepare and present a paper on your topic of expertise, can mean an assurance of that next department promotion.

Access to markets: If I were to build a digital communications device that every mining company needed, I would be seeking a relationship with a supplier of noncompeting digital equipment to mining companies, as a suitable investor/partner. If they have hundreds of customers with a need for what I have built and a strong relationship with the third party I am talking to (as their supplier) I have a very good chance of jump-starting my sales with their client list and they can mitigate all of their risk with an assurance of instant sales, by feeding my product into their existing channel.

Access to skilled people: If you have to recruit skilled scientist or engineers, they might not be attracted to betting their career on a start-up. It is sometimes easier to collaborate with a firm that has these skills on-staff. If they are not directly engaged with you, you can more readily absorb the cost of delays in project development and approvals, given that you don't have monthly salaries which will burden your project.

Community recognition: I recently had a project involving a medical practice in a remote part of Australia (a territory twice the size of Texas) seeking to place medical imaging equipment which could save patients a flight and 2-3 days of accommodation, to get medical imaging services when required. The investor offered to fund this as a convertible note (with a healthy interest rate) on the provision that this investor could announce to that community that they financed this life-saving equipment, to benefit the people of this region.

Being elected to the Board of a Company with someone famous: I had a real estate project which was moving to a public listing, in which the investor's key reason to invest, was to serve on the Board with the Chairman, who was well-known in Australia property fund circles.

The opportunity to socialise with the other party; Just the same as people take VIP seats at major business event to get a photo opportunity with the main act, people also will invest to have someone famous or interesting in their rolodex.

Migration: Some of the business migrants coming into Australia have to invest up to $2,500,000 into a business venture and then plan an active role in a senior management position for at least 2 years. If the position they nominate for is export marketing, they can then return to their home city and start the marketing process, thereby buying themselves several years of dual residence.

A significant first order: I was recently involved with a project in which the most prominent investor was offering an order totalling $10m of the product, in exchange for a sizeable percentage of the company. This enabled the project to have their development and production underwritten, as they commercialized.

Regardless of whether you are looking for investors, collaborators, or other innovation that could complement your project, you could well ask yourself what currencies would the other party value more than cash. At the same time, you should determine if the things you are going to spend the cash (you are raising) on, can be provided by the other party it will be cheaper and far less risk for both parties, to minimise the cash component of the transaction.

Understanding Risk

Risk means many things to many different organisations. The concept of risk analysis for intellectual property is all about the likelihood of achieving everything claimed by the principals. To mitigate or manage his risk, the Cornerstone Investor will consider the issues above and also these additional five:

Competitive Advantage: Investors will want to see that once the product or service is built ready for commercialisation, it will be able to deliver a competitive advantage over all other substitutes, as well as deliver this advantage over a sustainable period of time, such as the life of a patent.

Leverage: Most experienced Cornerstone Investors will look for how this product or process will deliver an advantage in the current business in reducing costs or increasing margins, sales reach, or average sale value. Typically the investor would like to know that one plus one *can* equal five in their existing business operations. This makes investment in the innovation more urgent.

Exposure: Any astute investor in intellectual property will understand that most projects, particularly those with inexperienced principals, will generally run out of money before reaching their commercialisation objectives. The investor needs to know the long-term exposure, particularly given that if he does not support his initial investment you could lose the project to a third party, who may pick up his additional rights at a small price or no cost.

Project Management: Professional investors will look at the qualifications and experience of the project team. Experience means specific work on this project, on other similar projects, on projects that have matured past this process and into commercialisation and team members' previous work environment. Rightly or wrongly, there may be a perception that if a senior team member has been an academic for all his or her adult life, the academic probably does not have the commercial discipline required to meet stringent deadlines and within strict budgets.

There is nothing worse than handing access to a bank account of $1 million earmarked for research and development to a person who has never seen that amount of money before. I recall such an event some ten years ago where the principal of the R&D firm immediately went out and ordered a brand-new Toyota Land Cruiser because he felt he had the budget. Being good at project management does not mean that any individual is good at budgeting, administration, or even commercialisation.

Investors will also look at how well the team members get along with each other, watching out for any friction between key personnel. Investors may invite parties out to dinner and change up the arrangements or where people stand at the bar before or after the dinner. Sometimes parties cluster together, and opposing groups will become evident in two or more social gatherings, gatherings often scheduled on the same day. Observing where people stand and who they talk to, investors will see whether parties mix and cooperate well or whether there is a rift.

Ownership: In some cases, the project idea has transitioned to a concept through debt funding by the principals and/or their families, creating a long and confusing line of intellectual property ownership. These families may have an expectation that when the Cornerstone Investor arrives, they will be repaid their loans with interest. In some cases, they then expect to retain an equity position in the project because of promises made to them by the intellectual property owners or principles. Cornerstone Investors see this as a red flag for protracted difficulties in ownership and progressing projects through the next and subsequent stages of development. In some cases, the principals may have the intellectual property in their name and sign a commercialisation agreement with the development company. This essentially means that any Cornerstone Investor, putting money into the development company, will have no intellectual property rights associated with his investment.

These must-haves and the associated risks to be considered in the proposal are essential in the first stage of early-stage commercialisation, when the project is exiting the research and development phase and now requires funding to reach the cash flow positive stage of its life cycle.

If you are able to add these elements to your information memorandum (the document you prepare to raise capital from sophisticated investors), then your project should be attractive enough to professional and experienced investors to get the funding you need. This chapter has not focused on the products, the markets, the industry or the people. Instead, this chapter has given you the tools to overcome the most common issues that leave most quality projects un-funded. If you wish to learn more, visit www. MakeMyInnovationHappen.com.

How I make investors

Without exception, most people enter my IP commercialisation program, want to be immediately introduced to potential investors they can pitch their same message to, in the hope that one of these new prospects will give their project the investment money required. Some of them even offer me a commission on the funding and it surprises them that I suggest no matter their project, that sort of investment will never happen.

I don't consider myself any better than the next guy in finding investors. I don't actively search for new investors these days, but still get most projects funded. The reason for this is that I don't find investors - I make them.

The majority of cornerstone investors who participate with projects in my IP commercialisation coaching program, are past inventors who have completed the program and have been able to take a profit from their projects at their value realisation point. These guys have done the commercialisation process from "proof of concept" through to "value realisation" and are now contemplating their next project. Most of them have also done the hard times in their original project (from concept definition, through research and development) and recognised just how hard and unrewarding this phase of the projects can be.

It is fairly easy for them to understand that they can make more money, on a shorter commercialisation cycle, by walking into projects at the post-proof of concept stage. In most cases, they can bring fresh commercialisation experience relevant to the industry that this new project is targeting and are happy to contribute $500,000 for a 20-50% equity position in the project, if their skill set can add massive value in the commercialisation process. In most cases, inventors appreciate the value of the experience and will increase the amount of equity offered, to encourage these Cornerstone Investors to partner with them.

The most successful cornerstone investors are the inventors who have closed out their project successfully and are now looking to accelerate another project in half the time or less. By adding their skills, their money and their contacts, they are able to achieve up to a fourfold increase in productivity, once they eliminate the learning and focus on where everything matters. Their experience makes them the perfect investor and commercialisation partner for most IP projects, as they play a mentoring role as well as looking after the business rules. They also understand the value of setting up the exit early, even as they start the project development process.

Having this know-how and industry contacts attached to the funding, is the hidden bonus that most inventors don't see, when looking for investors. This process works remarkably well because it reduces the risk for all parties and the commercialisation experience of the investor (as past inventors) tends to lend itself very well to project management on consecutive projects.

It's not hard to convince inventors who have successfully closed out projects, to become cornerstone investors with some of their dividends. They understand the rules, they have an affinity with the inventors and they also understand what needs to get done - and when. If they start from scratch with an idea, it could take them up to 2 years to get to the early-stage commercialisation and another two years to get the project to an exit. By investing in a post R&D project as a cornerstone investor, they reduce the two years of commercialisation to sometimes just six months.

The hidden benefits for inventors include more than just the know-how. Usually these cornerstone investors have a rolodex of contacts that are relevant to the industry or market upon which the invention is focused. There industry knowledge and process knowledge of commercialisation, reduces the risk associated with longer term development. Most cornerstone investors have experienced the pain of delays and will manage the burn rate very carefully during the commercialisation process, to ensure these delays are minimised but monthly outgoings are significantly reduced. In the case of unexpected delays, an experienced cornerstone investor will stop the spend rate immediately and suspend the project until the delays are cleared.

Most experience cornerstone investors understand that the amount of cash required is not required at once and this gives them the opportunity to manage the investment allocation over 6 to 12 months. They also understand the value of non-cash contribution, including services, access to equipment, facilities, large supply contracts and other contribution. They generally have a good understanding of government grants and/or tax offsets, which can be shared with their new project to reduce everybody's risk, without affecting the contribution level required to the project.

In summary, whenever I am asked can I bring investors to a project, I generally will say no. This is not because I can't introduce the right project to the right investor for the right outcome, but rather most of the projects that I am likely to look at initially, need a lot of work before they're ready for putting before seasoned cornerstone investors. These investors are generally looking for hands-on roles and have worked with me in the past. They know that I would not present a half-baked project to them, as they have experienced my requirements first-hand with their original project.

Once a project is accepted into the IP commercialisation coaching program, we first get to work on preparing the project for the cornerstone investor and overcoming all of the roadblocks before investment. Once all the requirements of an investment-ready project have been met, I open up the opportunities to our past program members, for consideration as cornerstone investors.

Not all of the successful project owners return as cornerstone investors. However, many of these will return many times, as they get more proficient at closing out each project and starting new ones. I have cornerstone investors asking to be flagged for projects in certain topics or market segments, where they have expertise. Others are constantly looking for opportunities to aggregate with their current project, to achieve a greater value for trade sale or public listing.

In most cases, everybody wins. With higher trade-sale or licensing outcomes, reduced commercialisation lead-times, an immediate injection of expertise and experience, walk-in sales and industry contacts, all helps to deliver a stronger return for both parties and answers the hardest question for most inventors – who will fund my project?

CHAPTER 5

Finance

"You cannot bore people into buying your product; you can only interest them in buying it."

—David Ogilvy

Allocation of Capital

Before you plan how much you need to raise and how you will go about it, you first need to be certain about what you need to raise the funding for. One of the most expensive mistakes inventors make is to focus on

Raising the Capital

There are four common investment stages in most intellectual property commercialisation programs: start-up, seed capital, venture capital, and IPO. These occur in specific phases of a project. It is important to understand the investment needs and tastes of seed capital investors and the differences between them and venture capitalists that provide substantially more money but at a more mature interval.

Understanding the requirements and timing of capital will help you understand who would or should invest in your project. During the first phase, commonly called start-up, we expect your project to receive funding from family and friends only. This is primarily because your project cannot yet show tangible value to prospective investors, despite that you may have a provisional patent registered. At this point in your project, risk far outweighs reward. People are investing in your personal capability rather than your idea or concept.

The second phase of investment, commonly referred to as seed capital, may require you to present your project to people you have never met before and who do not know your capabilities. The best way to get this type of investor to make a positive decision in your favour is to identify the ideal profile for you to present your proposition—and hopefully jump on board with funding your idea.

This chapter focuses on that second stage of raising capital and the profile of the individual or entity that may invest in your idea. Most inventors read American innovation books and textbooks and conclude that likely investors would be venture capital or private equity firms, but that is not the case. Venture capital comes in at the third level—after a project has shown steady, consistently growing cash flow and can prove that an injection of additional capital will offer four to ten times multiplier on current activity. We will discuss venture capital and private equity in later chapters.

In any practical model of seed capital investment, any post proof of concept project will be most attractive to a Cornerstone Investor. Typically, a Cornerstone Investor is a person or entity committed to the industry you are targeting that can benefit from investing in the project through a sustainable competitive advantage without compromising or restricting the distribution of the finished product. In most cases, the investment returns are a distant second to the competitive advantage and even to the risk profile of the project. Understanding that a Cornerstone Investor can offer more than just funds is the first step to this reciprocal leverage that comes with engaging with a funded partner or Cornerstone Investor.

In some cases, the Cornerstone Investor will provide capital for the commercialisation process, but not directly to the project. There may be a series of market validation or other tasks that can be catalogued as R&D that may attract taxation incentives in whatever jurisdiction the project operates. Under a collaborative research partnership, the Cornerstone Investor can pay these accounts directly (rather than into the project) and participate in the R&D function, thereby attracting the research incentives as a substantial bonus to the investment. In many cases, R&D tax concessions are of little use to a new company because they do not represent cash but instead a discount of taxable income in a year where the taxable income may be zero. Any savvy accountant will be able to establish the operating rules for a collaborative research partnership, which suits the taxation regulations in your jurisdiction.

Under these collaborative research partnerships, the Cornerstone Investor need only contribute the funds as required, thereby reducing his need for lump-sum cash. It's important for those invested in the project to understand that equity is released in direct proportion and timing to the payment of accounts or funding received.

I once witnessed a project where the collaborative research partner was promised a substantial amount of shares for performing some R&D in plastics but sabotaged the final listing of the company as soon as he received his advance shares. This crippled the intended public listing and resulted in nearly six years of litigation. Most of the pre-IPO shareholders may have lost their investments as a result.

Ticking the Boxes

Even if a Cornerstone Investor is not particularly sophisticated or experienced, he or she will require certain boxes to be ticked before approving any investment. If these minimum requirements are not met, anyone he takes your proposal to for advice will advise against investment. Your business case must include acknowledgement that:

1. The project team members are the sole owners of the intellectual property and that the person representing the project can act on behalf the project in making decisions.

2. The project is not party to license, lease, or other agreement to use the intellectual property.

3. To the best knowledge of the innovators, the project does not interfere with intellectual property rights of any third-party.

4. There are no actions, suits, or proceedings pending or threatened against the project claiming that it is infringing upon the intellectual property rights of others.

5. To the best knowledge of the innovators, no other party has infringed on the project's intellectual property rights to date.

Next, each investor will need to know specific information about the project, its personnel, the market, the industry, and the exit plan. Each investor profile will have its own unique set of questions for each of these elements, but they will generally be focused around the following items.

The Project

Given that the primary reason for your presentation to a prospective investor is to present your product, you should generally lead with your project scope. This does not mean you should talk about what a wonderful widget you are creating or how clever your team is for developing it. You should communicate what your project will provide to people who need it and when. Elements of the project part of your pitch should include answers to these questions:

- *What idea do you have and how long have you been working on it?* Investors need to know that your idea wasn't conceived yesterday, but also that it has not been under development for 15 years. They'll also want to experience long each member of your team has been working with the project, to ensure continuity and commitment.

- *What does your company intend to do for your customers?* This is the business case for how your project will make money. Why will people buy it? Why will they choose your product over a competitor's? And most importantly, how much money or time will your product make or save your customers—and your investors?

- *What is unique about your idea?* What sustainable competitive advantage does your innovation offer the buyers and end customers? Investors want to know what advantages your idea has that sets you apart from your current competitors. This is the leverage that most Cornerstone Investors are looking for in order to justify investment, especially in an unproven technology or innovation.

- *What big problem does your idea solve?* In most cases, a prospective investor is already familiar with the industry you are looking to enter, so you should also have a good understanding of the industry's needs and frustrations. You should be able to present a business case for why relevant decision makers need your solution to a problem they readily acknowledge. This same business case must reveal to the prospective investor a compelling argument as to why he or she should invest in order to be the person or company that fills this industry need.

- *How big can this grow given the right amount of resources?* This question is what is commonly referred to as the blue sky component of the business case or pitch. Having a sure-fire money machine is great, but having even a slight opportunity for some massive leverage in the future is always going to provide a more immediate positive decision by prospective investors.

The Team

Once a potential investor understands what the proposed product or service is, he or she will want to focus on the people who are developing it and will bring it to market. Investors expect to see a difference in these two teams because the skill set required for research and development is different from the skill set needed to commercialise. Your goal in the team section of your pitch is to communicate who team members are, why their experience and skills are critical, and how they feel about letting go of the technology once it progresses to a point beyond their experience and capabilities. Key areas to be covered in this part of the pitch should include:

- ***Who are key team members and how much of the project do they own?*** Prospective investors want to see that those closest to the project are there because their skills are needed—not because they are related to one of the founders. They want to know what each team member's commitment is going forward. This may include current ownership and commitment to acquire more equity in the project, such as shares in lieu of salary.

- ***What are their experiences and qualifications in this technology and past projects?*** Having a track record in developing intellectual property is a significant advantage but is not essential. We once worked with a project that had developed significant competitive advantages in carbon capture, which was developed by a very astute self-taught engineer whose only qualifications were in farming. The innovator was creative and dedicated, but we could see that the project might suffer at the presentation stage if he did not have support from more qualified personnel. His first partner was a physicist with 25 years of experience in systems process engineering. The partner had the practical experience and the science knowledge to explain how the project was progressing and why it was taking the direction it was. The two partners worked well together and eventually collaborated on other projects that transitioned into the commercialisation process. Although neither had the development or presentation skills needed to get this project commercialised on their own, together they were unstoppable. Together, they took the project international.

- ***How available are key team members to complete milestones?*** It is assumed that most research and development projects that are not subsets of major corporations or Universities are led by part-time innovators who have full-time or at least part-time incomes elsewhere. This can present a problem for a project moving forward, especially if there is a long lead-time and salaries are not offered. For most investors, it is more prudent to offer additional equity in a project than to offer salaries to the principals. Where key personnel have other commitments, they must be able to prove their availability to meet particular deadlines and milestones as and when required.

- ***How will individuals feel if the project does not require them anymore?*** Like humans, every project either grows or dies. At a certain point, a project may not require all the people who had a hand in developing it. A prospective investor will want to know if developers are reluctant to hand their project off to other parties. There are many projects that did not receive funding because the principals didn't trust that other parties could do a better job than they could and tried to retain control of the project, even if it had grown past his or her skill set. Most astute IP investors will recognise this and will turn the opportunity to invest down, no matter how good the opportunity may look.

- ***What human resources are you short of now and in the near future?*** In my first meeting with almost all IP owners who present me with commercialisation opportunities, I tell them, "What got you here will not get you where you need to be." Some people understand this and others do not. A different skill set is generally required to transition a project into a commercial environment. You may not have the skill set that's needed in the future, and it is not fair to ask an IP investor to pay for your learning curve. Investors will want to know that you are willing to listen to other professional opinions and that you are able to put the project into the hands of someone more qualified if necessary. It is easier to fund a leader with the right skill set, usually with a combination of equity and income, than to fund the learning curve of a founder who is reluctant to give up the reins.

- *How will you handle your personnel needs when you scale the project?* This question is just as much about ego as it is about personnel management. As you leverage a project up to a national or global scale, you'll need to be as economical with growth funding as possible so the investor and yourself retain more equity in the project. Are you open to help from the Cornerstone Investor? In some cases, you have selected that person or company because of their industry experience. This is not to say that you need to surrender your project to a third party who invests in it and then controls it. Investors will want to know that you have people or companies in mind for the commercialisation process of the project and that you may have already had preliminary discussions with these parties.

- *How critical are the top three equity owners or salary earners in the project?* In the case of early-stage commercialisation projects that are seeking Cornerstone Investors, there may be no one within the project who is earning a salary. However, if the project is already earning income and some of these distributed earnings go to salaries for key personnel, investors will want to experience critical those positions are, how committed those individuals are to the project, and what the damage would be if they exited the project early. You will need to have your responses to this ready but not necessarily contained within the presentation itself.

The Market

- *What does your project do for the market decision makers? How much will it save or make them in time and money?* These questions do not need comprehensive responses. It is more important to give a short five to ten-word summary of what you are creating and what need your product fills. Why would people switch to your product or service? What value would a buyer receive as a result of switching?

- ***What do they currently use and who supplies it?*** It is important to understand how loyal potential buyers are to their existing supplier. How difficult will it be to get them to switch? With several projects in the fast food industry, I have seen how difficult it is to specify a supply line into a major fast food company, because of the damage any threat to quality can result in both the brand and the loyalty long-term. You will need to know the current suppliers for the customer you are targeting and have a compelling case for why the current customer would consider switching.

- ***How will you get your message to the market?*** If you know your competitive advantage, how will you communicate that to prospective buyers? If you were to advertise on television, how would you pay for it? How can you justify the need for public promotion if you do not have a network of suppliers who can deliver your product to the people who want it? Do you have infrastructure and inventory to meet rapid market demand, if publicity should deliver prospects to you?

- ***How do the target customers currently solve the problem that you will solve?*** Investors want to see that you've examined all aspects of substitutability, when it comes to how prospective buyers solve that particular problem today.

- ***What is the actual market size for your region and beyond?*** Market size relates to the potential sales you could obtain within a given period in any given geographic territory. Be prepared to justify any numbers you present as market potential. Most investment pitches fall down at this point because the innovator has not invested sufficient time to quantify the market and instead relies on gut feelings to make projections.

- ***How will you gain some of this market share within two to five years?*** Investors want to see what your strategies are for securing some of the market in a given territory. They will be particularly interested in your expectations given the limited budget you will have against your competitors. I recently commercialised a phone-based application and we recognised that the market was for business people who had used their passport within the last 60 days or had moved more than 250 miles within the last two years. Given this demographic, we were able to fashion a marketing program using American Express and Diners Club to offer this app to their members as a free reward. The first mail-out opportunity was expensive but would deliver a free sample of our product to more than 11 million prospective customers around the world within 14 days. For that level of immediate market penetration, it would have been cheap at twice the price.

- *What do you need in place to make this happen?* To give any investor confidence in your project, you must have an answer to this question. Your response should include a plan, a budget, and a list of the people needed to implement the plan, as well as the product that has been market tested at the recommended price.

The Industry

- **Who are your competitors?** Your industry consists of any company that provides any product or service that could serve as a substitute to what you will offer. You may have a competitor in an entirely unrelated field, but if it solves the same problem then you must consider it your competition and assess the cost for your buyers to switch from that product to yours.
- **What gives you a competitive advantage?** What sets your product apart from the competition? And more importantly, how sustainable is your competitive advantage? This answer is sometimes called a unique selling proposition. Investors want to know what sets you apart and how that difference can be maintained and controlled by you.

- **What advantages does your competition have over you?**
  You might think it's better to not disclose this to investors,
  but that's not the case. If you do not disclose what you know
  about the competition, your investors will find out eventually
  and may feel you have been dishonest with them or are naïve
  to the competition. Either option could result in you not
  getting the funding you need.

- **Compared to the competition, how do you compete with
  respect to price, features, and performance?** We recommend
  creating a matrix of features and benefits that shows what you
  have and do and what your competitors have and do. When
  producing a matrix of this nature, you should only detail the
  features and benefits you presently have. If there are features
  and benefits you plan to roll out in the future, you should
  not disclose them as available today. This matrix can provide
  excellent PowerPoint slides for presentations and work well as
  appendices in offer documents and business plans.

- ***What are the barriers to entry?*** Are your competitive
  advantages protected by patents? Can you prevent others from
  entering the market by some exclusivity to resources? How
  would you go about protecting or defending your market
  share from new market entrants? It could be that you have
  no barriers to entry to prevent other market players, and
  most businesses would understand this. In commercialisation
  of intellectual property, there must be some competitive
  advantages. The more sustainable these are, the more they will
  serve to become barriers to entry. If you present a patent as a
  barrier to entry, investors will want to see that the patent has
  been converted to a full patient and not just a provisional one.

- ***Who are the current players in your region?*** If you are a very
  small business, your competitors may only be local retailers
  or service providers. If you are manufacturing a product, you
  may have competitors in other countries with no interest in
  pursuing what you do your corner of earth. Either way, you
  must list them along with their expected response to your
  project and the risk management strategies you could put in
  place to protect your business.

- ***Who are the players on the global stage?*** Investors want to know that you've thought about growing your business beyond your city and perhaps even your region. You should not be concerned with manufacturing representatives or agencies of major international companies but rather on the international players and suppliers to those agents.

- ***How will your competitors respond to what you are doing?*** Will competitors come hard with massive advertising budgets that prevent you from getting a market foothold? Are they likely to make an offer so they can shelve your product and reduce their competition? Could they perceive this as better than the product they currently have or as complementary to their range of products? In other words, could they be a potential trade sale buyer who could offer you shares in their listed company in exchange for shares in your private company at a later date?

- ***Who would be willing to buy the technology once it is developed?*** Investors want to see that you've spent some time thinking about what companies might be suited to buying access to this innovation, in the future. One of the key issues that sets successful business owners apart from others is the ones who have built a business for the second consecutive time, nearly always make significantly more money from the operation and the sale. This is because they go into planning the business with an end buy-out in mind. Successful owners build their business knowing that it will one day be attractive to suitors who may be competitors at the present time. Astute investors invest in those who structure their business or innovation to maximize trade sale price at any given moment.

- ***How would you feel about someone buying your company in the future?*** This question is generally one that gets asked verbally. Investors will not expect you to answer this in writing; they just want to experience emotionally attached you are to the product concept. To them, a business is an asset and assets may need to be traded to make a profit. In most circumstances, Cornerstone Investors will be looking for innovators with the same business attitude.

The Exit

- *At what point can I start getting a return?* Investors want to make a profit on their money. No surprises there. However, some have different investment profiles than others and may require an income stream, or the knowledge that the company is receiving an income stream, which in turn will mitigate the risk going forward. Some innovators will want to receive their investment in shares of a publicly-listed company. They would be less concerned about which entity these shares are from, be it the vended company or the buyer.

- *To whom can I sell my shares and when?* A minority shareholding in a private company is the worst type of investment. Investors want you to show them how they will realise cash for what they put in – such as tradable equities or cash from a willing buyer. In some cases they may offer you convertible notes, which may be kept as debt until the shares are listed. This will give them an opportunity to withdraw their support if you do not meet deadlines or development milestones.

- *How will you protect my investment?* Every investor needs to manage his or her risk in every project, all the time. If you are able to provide a long-term risk management strategy, the investor will have confidence in your ability to protect his investment—and yours.

At various stages of the investment process, many investors will have different priorities on this information. Some may require extensive detail for certain aspects of the above information, so it is always prudent to have a due diligence file—one that includes supporting documentation of all the answers to these questions— available for inspection by any investor. This will save you hundreds of hours of due diligence, as each investor qualifies the opportunity for their investment profile and risk profile.

When calculating value for prospective investors, always start with earnings. Make sure you are focused on what the project can provide in net income. Traditionally, your value is not in your patent or how clever your idea is. Your value is found in the net earnings the project can generate—even from the early stages—through a sustained period. Having a patent means you can generate this volume of earnings for a sustainable period on an exclusive basis. It does not mean that others cannot come up with ideas that are similar to yours. If your potential earnings are not substantial, your patent is not worth zip.

CHAPTER 6

Getting Traction

"Don't worry about failure; you only have to be right once".

—Drew Houston, Dropbox

One of the most challenging roadblocks for innovation projects is getting funded. Most people believe it is because there is not much venture capital around for investment in innovation, but the truth is quite different.

Imagine you are head of a venture capital company or private equity fund. You have just purchased a large national chain of fast food outlets, which is the number three in this industry. As the two in front of you are global franchises, you are less concerned with these market positions and more concerned with generating more profits for less effort.

The three things your shareholders are going to want to see from any investment you make, is (1) a reduction in costs, (2) some blue sky (new innovation which can generate more sales, make the products stand out or reduce operating costs), and (3) an exit (which we refer to as a value realisation point). In approximately half of the cases, a private equity fund would want to have a public listing as the exit point, so the more of these three elements, the greater their arbitrage. It makes sense for them to shop around for intellectual property, which can reduce their costs, increase their margins and customer base and also help to differentiate them from the pack.

In the above case, a $700m company was prepared to enter a collaborative research partnership to develop products and processes that would differentiate their fast food offers. They invested nearly a quarter of a million dollars in development and testing, just so they could get first rights to the product, in their markets. This makes perfect sense, does not it? They locked in an 8-year supply agreement and helped get the product tested and approved, so they could get exclusive boasting rights to their market.

So, how do they select the innovation they want to include in their IPO offer? There is, of course, no set general answer to this question, but this Chapter presents the top five must-haves, to get most innovation project funded.

The Business Model

So you prepared that killer business proposal, you got interest from the right profile of investors, and now you are looking to make the presentation and get the funding. What about your business model? Many innovators focus on the operations model or the method the widget performs, but more than half will forget the underlying business model. Investors need to see evidence of how this widget and the associated intellectual property will turn into cash for the company and its investors.

The business model does not have to be complex or convoluted. In fact, the easier the business model is to understand, the more likely it is to get funded. Some business models may be as simple as "we buy seeds, we grow trees, we sell seedlings, and we sell fruit." The investors will want to see where you grow those seedlings how you incubate them how you water them and who will buy the seedlings and fruit at a later time. You will need contracts or offers confirming that there is demand for this fruit and for the seedlings, if you are to commit somebody to put their funds into your project, you have a handle on your underlying business model.

Several years ago, I had a project which built a machine that would capture 99.6% of carbons, nitrates, and sulphates, as well as particulates, from a coal-fired power station. This had a unique business model, in as much as when we tested the market for coal-fired power stations, we understood that most operators would not spend money to clean up the air, as they considered this an expense that could not be recuperated. This product was designed specifically for the carbon credits schemes being planned and offered in Europe, Asia and the Americas.

The concept was that the facilities management company (and we had one in mind), would take a licence to produce these carbon capture devices, and instead of selling these to their power station clients, would offer to install these, clean up the area around power station, and take the carbon credits for themselves as remuneration. With the number of coal-fired power stations in those three markets, and the knowledge that more than one power station driven by coal, is being commissioned every day of every year in China alone, the developers knew that their model would be far more acceptable, than building and selling units, to coal-fired power stations.

The project development was stalled somewhat, when the governments of Australia and Canada review their commitment to carbon capture or storage, and the American Republicans gained control of their upper and lower house, forcing environmental issues off their agenda. This didn't kill the project. It merely put the project on hold, until the relevant parties were again discussing legislation, relating to clean air and carbon credits.

This was a very complex business model, but was easily explained to economists and scientists working in this field. This is not to say that every business model should be as complex. The more simplistic the business model, the easier it is for the Cornerstone Investor to understand and more importantly, the easier it is for that Cornerstone Investor to explain it to his investment advisers, and other potential partners he would like to bring in, to share his risk.

Ownership Structure

R&D Projects need to have the appropriate accountabilities by the managers, so having the right corporate structure makes the directors accountable to the shareholders, in most jurisdictions. In Australia, we recommend a public, unlisted entity, which will require a public unlisted company, with three directors, annual audited accounts and an independently-maintained share register. The issues associated with this configuration, have been explained in earlier chapters.

In larger projects, some investors will try to leverage this information, to better themselves against the others, creating a dynamic that ultimately implodes the project. I was recently involved with a project that covered three different countries, had three different corporate structures in these three different jurisdictions, and had raised money at different prices, from different investors. Before we could get traction with further investment, we had to restructure the entire entity, with a single structure into a single domicile, so a public listing could become a reality.

This took nearly ten months to restructure, as minor parties jostled for advantage and some parties became obstructive for leverage. This extended the burn rate for the commercialisation process and although the project had its first eight-figure sale, the rollout was approximately two years late.

Exit in Mind

If anybody is going to jump into your project, they firstly need to understand how and when they can jump out. Without a clearly-defined exit, their advisors will tell them to buy public stocks, which can be traded at any time. There is no set rule here, but depending on where (in the lifecycle of the project) or the size of the reward the project can offer a six to 24 months exit to a Cornerstone Investor.

The exit point does not need to require any or all parties to dispose of their existing equity. A better description for exit could be of a value realisation point. There has to be a defined interval in the development process, or the commercialisation process, where values can be determined absolutely and parties can buy or sell shares or other securities within the project. This does not necessarily mean a public listing.

I was recently involved in the aggregation of a group of property rent rolls, spread across Australia, which were brought together for a public listing. The commercial intent for each of the individual owners was to provide an exit point as a public listing on the Australian stock exchange. Where no individual property manager company was large enough to list in its own right, the aggregation of all of these made perfect sense. Given that a private company has a rent-roll value of $2 per $ earned, a publicly-listed company can command a 6- ten times profit multiplier, which increases the equity value of each individual business owner, as a collective.

We were able to raise the seed capital to put the financial structure and corporate structure together, which then enabled us to bring all parties to a public listing under a suitable document. This met the outcomes of both the principal and the individual owners, as they were able to obtain a higher-multiple valuation, based on a public company against individual private companies.

The Right Team

Would you be comfortable with the cabin crew of an aircraft you are a passenger on, decided to fly that aircraft? They could be the best cabin crew on the planet, but that does not give them the skill set to manage a fully laden commercial aircraft. In many projects, the R&D team complete with their skills experience and attributes, are not necessarily the appropriate people to drive the project through the commercialisation process. Investors know this, and will always look to a formal handover, as a check to ensure that the project will never be cycled back into R&D, as soon as the first commercialisation roadblock is hit.

It will be critical to have a solid experienced commercialisation person or team to manage the project beyond the Proof of Concept. At the point of handover, referred to as an "internal trade sale audit," the project needs to be in the hands of a commercialisation team, as the R&D process has been declared complete by the current R&D team, at the funding presentation.

Even within a multinational Corporation, the R&D Department effectively conducts an internal trade sale audit, passing the project off to the commercialisation or marketing team, for the next phase of the project's development.

Most experienced investors and managers understand the risks associated with leaving a project in the hands of the inventor, for the entire commercialisation process. The two major risks are that the project will be cycled back into R&D to accommodate prospective clients wants, or that the project budget will have to factor in the learning curve of R&D personnel on the commercialisation process. Both issues can eat a commercialisation budget very quickly or kill a project by effectively removing it from commercial availability.

There have been many cases that I have witnessed, when teams have received funding before my engagement, that had run out of money because the project team had run into a commercialisation roadblock and their response was to push the project back into R&D to build an iteration to overcome the roadblocks. This will eventually cycle the project back into where the R&D team are most comfortable – the R&D process.

I recently reviewed a project which had an extremely good product, was receiving a lot of interest from government buyers and was already processing sales. The project needed capital to step the commercialisation process up a few notches. However, during the scoping interview, the principal made it known that he did not want to relinquish control of the project, or the company that was to own and drive the project. He also wanted to have this project reflect his achievement, so he could be recognised by his peers.

After careful questioning, I recognised that his intention was to drive the commercialisation process from within his research and development background. I could see the project would not achieve its intended goals, unless you have the right people to drive it. I passed on the project, which is sad because the product was an excellent product and I had friends who had already invested in the project and needed me to rescue them.

The Right Amount of Funding

If you have not prepared your budgets and researched your spend rates, or you haven't timed your patenting expenses and so on, most professional investors will walk away from the project. Worse still, some may fund you with the intention of running you out of money, so they can offer the additional finance at a much steeper rate, thereby taking your project.

The key here is to not ask for too much, but never ask for too little. There are some costs that can be pushed out to beyond the commercialisation process. These costs should be paid for out of future cash-flow. Because the value of money at an early stage of the project is far higher than later, the project team should only seek funding for essential items that are critical to bringing the project to a positive cash flow.

In most cases, it's better to present your projections in a statement of source and allocation of funds. This allows you to put your cash flow projections alongside your equity placement or debt placement as agreed. Investors can assess at one glance, how the cash flow will track over the life of the commercialisation process.

It is most imperative that money received (for either debt or equity) should carry the project into a positive cash flow situation. If, after the expenditure of these funds, your project is still incomplete, you will need to be extremely nice or ruthlessly tactical to dig your project out from under the chaos that will ensue.

One of the biggest issues in commercialisation is always budgeting. If a project is under-budgeted, the money will generally run out before the project is commercial and everybody's investment is at risk. If a project is over-budgeted, the foundation shareholders are diluting themselves beyond what they need to, thereby giving away more of the project than they should have.

About a year ago, I looked at avery promising project, by a group of software developers, which I believe was truly a global game-changer. I did some rough costings with these guys, so I could get an idea of how much they actually needed to drive the project to a commercial reality.

They did not engage me at that time, as one of the prospective investors objected to spending the money on an independent consultant. The project principals made an agreement with this Cornerstone Investor, for a commitment that was substantially less than what was required, under the investment calculations we have made. The project subsequently ran out of money and everybody was stuck.

The foundation shareholders had more than they could afford to lose, tied up in the project, and the Cornerstone Investor had a considerable amount of funds buried in there too. It wasn't until the parties were able to agree on a fixed valuation of the innovation and IP assets, could we prepare an offer document that would raise the appropriate amount of capital to reach a cash flow positive position. In VC (Venture Capital) jargon, everybody had to take a haircut, which essentially meant revaluing each of the shares issued, to accommodate the equity to be issued to new investors.

Focus on the Cash

In many cases, intellectual property may have a number of streams of cash flow through iterations of the product or service. A highly disciplined commercialisation manager will identify the shortest path to market of each of the products and focus on the one product with the shortest path to market, the least risk and the lowest budget, or a combination of all three. This micro modelling will reduce the risk for the entire project and shareholders, but not necessarily kill off other options.

The intent should be to have one project opportunity reach commercialisation, start earning the funds required to pay for the other products and/or services to be bought online. The Chinese call this "…to fish where the fish are." Others refer to this as the low hanging fruit. In most cases, one path to market could produce sufficient cash flow to generate the commercial activity for other paths to market and therefore reduce the amount of capital required.

Presenting this in your capital raising proposal will impress any investor, as it shows you are being prudent with the money required and you are not likely to dilute yourself or them, as a result of underestimating your budget requirements.

One feature in any investment proposal that is sure to gain favour with prospective investors, is the commitment that the principal or principles make, to deferring some or all of their earnings, until the project is profitable. This will give investors assurance that they are not funding a lifestyle while the project is being developed, but are rather funding the project to the shared outcomes when they and the principles are rewarded accordingly.

Risk/Reward

There would be two separate values on an investment if one was to deliver the return in six months and the other was to deliver the return in 24 months. If the amount is the same, the second investment is only one quarter of the value of the first.

In most of our commercialisation projects, we look to a Cornerstone Investor who has a commitment to the industry and/or the markets, but we would only ever place them into commercialisation project for a period of six months to 18 months, depending on the project and the returns.

Most successful Cornerstone Investors are also looking for leverage, for their own projects or for market advantage in some way. They may be a buyer, user or component supplier to your project, and will stand to benefit from additional sales for every unit you produce and distribute. They will want to limit their investment period but extend their engagement as shareholders. In some cases, it is prudent to offer a measure of exclusivity for a limited period, as a sweetener for the investor.

It is very difficult for a commercialisation team to present their project, unless they are absolutely certain that the rewards and risks have been calculated and assessed. In any prospectus or other public capital document, this due diligence process is conducted by independent parties. In early-stage commercialisation projects, a lot of the claims are not independently substantiated and so investors will want evidence of any claim relating to risks and returns.

There are many instances of project principals who have not accurately calculated the lead-times in getting a project to market. Any smart project management team will have planned out all resources, including the timing of those resources, for costing and lead-times.

An astute investor will look for how the project is presented logistically. This will tell most people how thorough the developers are, in how they allocate resources and deal with the natural bottlenecks that occur in any project.

I recall an aggregation project in which we were set four months to pull 12 parties together for a public listing several years ago. After the first phase was complete, the principal got ambitious and decided to include another very large entity into the mix, before the listing. The due diligence alone on this task, created an additional three-month lag for all the original principles, who had become foundation shareholders in the project.

I was approached by one of these foundation shareholders, because his business was now tied up in the pre-float due diligence process, while his wife was battling cancer. He was not able to retire from the business to look after her. I was powerless to intervene, as the managing director of that merged project, took it upon himself to extend his offer to several other parties, in order to make a bigger impact on the financial markets, at listing.

This behaviour was called to account by the capital markets, when the IPO was announced. The target listing price was not achieved, as the markets viewed those delays as wasted money, given the time value of money to any investor. Ultimately, the additional value that was accrued as a result of these delays was not retained by the managing director, as the target share price after listing, was not achieved.

Just to summarise, these five critical must-haves are essential in the first part of the early-stage commercialisation process, when the project is moving away from the research and development phase, and requires funding to reach the cash flow positive stage of its life-cycle.

If you are able to add these five elements to your information memorandum (the document you prepare to raise capital from sophisticated investors) then your project should be attractive enough to professional and experienced investors, to get the funding you need. We have not focused on the products, the markets, the industry or the people, in these first roadblocks you will experience, until later in this book. Instead, we have tried to give you the tools to overcome the most common issues that leave most quality projects un-funded.

Timing and Sequencing

One of the most important tenets of investment and accounting is the time value of money. If for instance, $1 million is required to commercialise a particular intellectual property, there would be a very small chance that all that expenditure would be required in the first month. If the development program is over twelve months for instance, there would be a monthly budget required which matched to milestones in the development phase.

Given that these milestones were achieved, the funding for the next part of the commercialisation or development can then be drawn down. This will reduce the risk significantly for the Cornerstone Investor and make the investment far more attractive to a wider range of Cornerstone Investors.

Asking for $70,000 a month for 15 months, may be more affordable than asking for $1 million upfront, to some Cornerstone Investors, particularly those who manage their cash prudently from within an associated company that could later use the technology. Be mindful of only asking what you need, when you need it. The more attractive your project would be, because of the reduction in capital required at certain times, the quicker the decision from a wider range of Cornerstone Investors could be available to you.

Setting the Plan

Once you have your funding commitment in place, there is an expectation that you will make periodic draw-downs against the agreed budget. To justify these, you will need a very specific action plan to commit all of your resources to meeting the agreed milestones on the timetable you have promised.

In most cases, people set about building a business plan consisting of hundreds of pages. I would consider this only necessary if you were attempting to borrow money from the bank, which will require the evidence of your project's capacity to repay. Instead, I have always preferred an action list of tasks, sequenced to each milestone and each allocated to one person, who becomes the person accountable for that task.

In most cases any particular task may require the efforts of several personnel but it is more prudent to have one person in charge of the accountability of each task. This allows macro management of a project through weekly or monthly project status review meetings, where you only look at milestones, tasks completed and any upcoming issues. If everyone knows what they're doing there is less likelihood that people are going to slack off. If the project leader on a particular task is held to account for that task, he will drive the other people within that task to complete their parts in a timely fashion or become accountable to the PSR committee, at a later time.

Once you have an action plan, you need a budget. In some cases you will not receive the amount of money you pitched for. As a result, you may have to compromise some aspects of the development or commercialisation programs. Some commercialisation tasks may be left to a later time, so they can be paid for out of cash flow. In most cases, investors encourage early income to be re-invested into commercialisation, as a way of reducing the allocation of equity thereby diluting their equity and of course, yours.

Set Achievable Milestones

For every commercialisation process, milestones are the only way of measuring and calling to account all resources. As well as measuring progress and ensuring a project is on time and on budget, milestones are used as payment points, for where equity or salary is payable to personnel involved in the project. Once the milestones are achieved and the project status review team are able to tick these off as completed, the project should be able to allocate additional equity to those members who have elected to receive equity as remuneration for the services.

In almost all cases, I would not recommend that you pay salaries in the R&D or commercialisation process, unless it is from earnings. Any development or even early commercialisation tasks must be contract work. Nobody wants to see the project paying salaries before it earns sufficient income to pay for those. You effectively take away the incentive for people who are paying salaries to get the project to a break-even.

Milestones are also used to confirm for the investors, that the next tranche of investment funding is due and payable. If someone was to invest $1 million into a development or commercialisation program, it would not be prudent of them to place $1 million into a bank account and allow the company members to draw upon this at will. The acceptable method would be to allocate the resources against milestones and for the company to allocate the equivalent equity, as each milestone is achieved.

Setting unattainable milestones will become an encumbrance for the development team or the commercialisation team, depending where the project is. Clear measurables and timelines should be provided and agreed to by all parties as achievable and reflective of a true attainment of that milestone.

Allocate, Measure, Report

As you get your project started, all eyes will be upon you. It is important that you have planned to obtain project traction very quickly and that you can prove to everybody involved that your way forward will succeed. The way to do this is to allocate, measure and report. The assumption is that you have a technical advisory panel which does not report to you but that you must report to it on your progress and (in most cases) your funding will be in staged drawdowns, to be approved by the technical advisory committee.

The first part of this is to allocate your tasks and resources. It is important that all members of the team know what has to be done, that all of their tasks have been allocated to each of them and that everyone knows what they and their team members are responsible for. It is important that even though one particular task may have a bunch of people involved in it, there must be one person allocated to the accountability. This person is in charge of that particular task and must answer for it the task is incomplete

The Action Plan

There are many software packages for GANTT charts and action plans available on the web. Most of these can be adapted to suit any project, but possibly the most suitable method of allocating tasks is to use a plain spreadsheet. It is important that you are able to complete the task allocation on no more than a few sheets of paper and, if you do distribute it digitally, that all members of the team have access to the necessary software to read it.

This means that the common plan may appear very comprehensive in the early days, but may prove unworkable when nobody else on the team has access to the software package to read it and cannot afford to buy the tools to access it. It will be important for all members of the team to be able to edit this document and that any edits can be channelled back up to the master document for approval.

If you were to set up a single spreadsheet for a project task allocation, it is best to group tasks associated with shared outcomes. This way all members of the team can see who else has tasks on this outcome and indeed sharing other resources such as plant and equipment or ingredients and components, if these are required. The columns you will need will include the following:

- task group
- task
- person accountable
- resources required (includes people)
- prior tasks that must be completed before
- start date
- completion date
- other resources

You can download a sample of this form from our private member's area on our dedicated website www.billion-dollar-napkin.com.

The first milestone that you list on any action plan, which must feature all names as the person accountable, is the end goal. Each of your project status review meetings will need to have this end goal read out or understood by the team, before tasks are examined. It is important to measure everything against this end goal, because some of the tasks may prove irrelevant at a later stage, as events unfold.

It is also important to understand that the end goal is still achievable at all points during the project. There may be times when the end goal does not seem possible with the current intellectual property you have, so the project status review committee may elect to have a party research additional technology or opportunities that could plug into what you have, to help it meet the end goal.

It is important that when the task list is handed out to everybody, this should be done in person and each person (preferably in a meeting of the project group) should be given ten to thirty minutes to assess the task list and determine that they can or cannot achieve it. You need to have this commitment before the team breaks up, or the project may end up being assessed as impossible, when two or more parties start to fall behind. When I project-manage an action plan I tend to tell everyone that we do not leave the room until we are absolutely certain that we can be accountable for all of those tasks and accept any consequences for not delivering, as individuals and as a team.

There could be cases where a project sets unrealistic goals and although most of the team members secretly feel it is unattainable, they might not vocalise this because of their loyalty to the project leader or the pressure they feel in the meeting, given that everyone else appears to agree to commit to this. I worked with an international team who were building a global App which had a very useful function but the R&D team had convinced the chairman of the company that the product was ready and the other members of the development team knew the product was far from ready. Being an international project there were a number of different cultures involved in the project and many of the team members did not want to speak out. As crazy as it sounds, they were prepared to commit themselves to long-term commercialisation of a product that was nowhere near commercial ready.

This type of groupthink can be very dangerous in research and development applications and can be even more fatal in the commercialisation process of any project. The implications of waste of resources can be lost on people who are hell-bent on protecting their own employment positions and who feel they can reallocate the blame to other parties, if the project does not work. Having a commercialisation master plan with a full task list and accountabilities ensures these people cannot hide. It took several outside parties and a change of project management, to get this project back online, albeit six month late.

This project was never going to be able to meet the original deadline for the commercialisation tasks, because the research and development was incomplete and there was no internal trade sale audit conducted to transition the product from R&D to commercialisation.

Measurement

Tracking the progress of the task list can be automated very easily. Firstly, you can set up a mail list which will email each member of the team when the task is about to begin and when the task is due for completion. These can be automated reminders which can be set up in advance, so the person accountable will have no reason to say they forgot.

It is always helpful to ask each accountable person to report to you with a quick email once a task is finished so that you can tick it off your list. The other important aspect of this is momentum. Sometimes, you will want to have a public display of your action list so the other members of the team can see that task being completed, as and when it falls due which will put additional pressure on them for accountability.

This individual accountability can be private and without pressure, so that if any member is struggling to achieve one or more tasks you can step in with additional resources and keep the project's momentum. There are times when some team members fail to perform as they should and if you have access to this early warning system, you can implement some positive responses to ensure your project will still meet the overall deadlines. One important aspect of this is that there are no surprises and that there is no loss of face for anyone who is not able to complete their tasks quickly.

Project Status Review

Before a project begins the development process and certainly before the commercialisation process, it would be prudent to appoint a project status review committee (mostly referred to as a Technical Advisory Panel – not Board) to approve the allocation of the funding going forward, as well as to measure the progress being made by those receiving the funding. Commonly, the project status review committee consists of personnel from research and development, from commercialisation and from the investor. I strive to have an independent chair on the project status review committee, so that milestones are not hijacked by any party or declared unachievable or unattainable by one party, to further their own self interests.

The most workable project status review committees consist of five people being one each from commercialisation and R&D, two from the funders or their nominee and finally, an independent chair. We generally convene these committees for 18 months with monthly meetings and allocate an equal equity contribution which is pro-rated to each of the 18 meetings that they attend. Sometimes the project status review committee meetings are set for much shorter intervals such as weekly instead of monthly.

There must be provision in the charter to remunerate regular attendees differently, if the meeting frequencies need to be changed. At no stage should this committee be referred to as a board, because of the legal implications of responsibility and accountabilities associated with board members. If any of the members are also members of the Board of Directors of either of the entities, their roles should be differentiated in the charter.

Commercialisation Point

For most innovators and project leaders, there is always the temptation to add extra bells and whistles to their project, when clients or buyers identify these as something they would like. The temptation is always there to open up the R&D expertise and start to re-develop the current version of the product, to incorporate these extra features. This is another fatal flaw in the commercialisation process and represents one of the six major roadblocks that you will face in any commercialisation project. The more development-orientated your management team is, the more likely you will not be able to stop them from pulling the project back into an R&D cycle.

The only time it is acceptable to add features to the current project, assuming that the current project is funded and must meet certain performance criteria and commercialisation deadlines, is if one prospective customer or more is prepared to pay for the development. In a recent project I have worked with, one of the major multinational prospective clients identified a feature he would like to have on the product, and the innovation team were prepared to stop production to incorporate this feature into the product. I stopped them.

Eventually, after much discussion, the agreement was to proceed with only those features that could be paid for out of the first client sale. It was important to emphasise to the development team that their core project must proceed as it was and any features that were to be incorporated into the new iteration, were to be fully paid-for by the client that ordered them - but owned by the project.

As a result of this meeting, we drew up a list of features that clients had asked for and we prioritised these into must-haves and nice-to-haves. We then costed out the development in time and resources, before we then looked for clients who could pay for these development sub-projects, as part of an extra feature on the version of the product.

It took several phone calls to major prospective clients, before we had locked in the top three enhancements, with forward orders to cover the development cost on these and an assurance that the current development team would not be bogged down or have resources diverted to accommodate these extras. It was decided that each of the top three enhancements would be outsourced, at a fixed fee, to ensure they were concurrent to the core development program.

With this type of discipline, the project will not suffer delays or worse still have the resources depleted before the core development is ready for sale to first round customers. The temptation for any development team is to take the product back to R&D and conduct further enhancements, because this is their core expertise. The requirement of the commercialisation team is to have the discipline not to allow any project to slip back into R&D, unless the current version or planned program becomes evident that it will not meet the development or marketing objectives.

Traps to Be Aware Of

In the commercialisation process of any project, there are common issues with which nearly every project will need to deal. Some of these are size dependent but most will occur regardless of how many members on the team. The following is a quick summary of what these issues could be and how to avoid them:

Groupthink

There has been much written about groupthink: the persuasion of a powerful alpha personality on a subservient group. This can drive the decision making underground and although dissent does not disappear, it is not vocalised when it should be. In extreme cases people will ignore the alarms and continue to work on their part of the project without drawing anybody's attention to the looming dangers. Groupthink also damages the creativity of any group as people stop to think independently and will no longer provide options and suggestions in meetings.

Where unrealistic deadlines have been imposed upon the group people become stressed and anxious but will not always speak their mind if the alpha personality is truly dominant. In some cases I can spot the groupthink situations, simply by assessing how many team members have been let go on a project before my arrival. If these have been let go in groups or in monotonous regularity over the development time, they could be frustration firings or sacrificial lambs all with the intention of driving other members to agree with the alpha personality.

This can burn resources and although the alpha personality will have a whole team to blame at the end of the commercialisation period when the project fails, he or she will still not have a product and will not have a commercialisation budget to acquire one. Groupthink can be very dangerous indeed and having a technical advisory panel and/or a project status review committee without that alpha personality, is a clever way to ensure that team members will speak up about urgent issues efficiently, regardless of the (less onerous) consequences.

Internal competition

In most cases, R&D teams compete for the common resources against each other and are judged on outcomes. This drives team members to be competitive and in some cases, become political. Some project managers like this air of competition as it adds urgency to the development process. However, this can be fatal in the commercialisation process, if teams are not working together as they should.

Some examples of this type of unhealthy competition includes hoarding of information, restricting access to resources and setting up exercises to waste the time of other teams, to provide a competitive advantage for the design group. Having friendly competition can be an advantage, but if that competition starts to get aggressive, it is time to switch the members of each team around so that people do not become adversarial in the development or commercialisation process.

The perfect example of this can be found from a culture that dates back 41,000 years in Australia. Our indigenous culture consists of clans and groups of nomadic families, who moved around constantly so they did not impact too heavily on any one particular area. To avoid aggressive competition, they would meet up with other groups from time to time. Instead of warring with them they used to swap teenagers by identifying members of the group who were not paired off but of eligible age. The elders would then swap boys between the groups so that they would enhance the bloodline but just as importantly, they would not be at war with each other, from that point on.

This is how sons and cousins became related across thousands of miles, in an environment where one group might not see another for two years. That worked well for 41,000 years and we can take lessons from that, to prevent communities from becoming adversarial.

Hoarding

As competition might start to heat up in early-stage commercialisation, some teams may hide resources or have those limited resources allocated to themselves, at the expense of other parties. The purpose would be to ensure that they do not run out or experience delays in accessing resources, when their collective resources are looking scarce.

Fostering a good team attitude across all the different cells in a commercialisation program will ensure that all parties will help others and the development program does not become an "us and them" attitude in the race to achieve the individual tasks at the expense of the overall goals.

Valuation

As mentioned in earlier chapters as the process of commercialisation begins, some sales will be initiated. The commercialisation team must be mindful that any outsider, who values the company from this point on, may use a historical sales perspective to project the future value of the company. This is patently wrong on so many levels, but is common practice for the average accountant who may be required to conduct due diligence on behalf of a Cornerstone Investor.

The fact that there is no promotion budget and/or sales team would not preclude these people from assessing that your six sales worth $100,000 over the past three months (in the early stages of the commercialisation process) should not demonstrate a project a value of $400,000 a year for the next four years. It may be prudent in some cases, to actually provide the products as field-trials, in the early-stage of the commercialisation process. The project could simply charge a service fee for the delivery and implementation of a particular product, where having the product in the field working will provide invaluable feedback for the team.

I once presented a project for a Cornerstone Investor who had become a very young and successful multimillionaire in Australia. As a relatively newly successful entrepreneur, this young man had surrounded himself with friends from his university days to engender loyalty and camaraderie. One of these individuals was a fellow graduate who became an accountant and whilst working in an accounting firm in the suburbs, had worked his way into giving advice on investments and projects for this entrepreneur.

Our presentation to the entrepreneur was for a pre-IPO project rollup investment, where we were to acquire a public listed shell and load the contents of the R&D project into the shell. We would then relist the project onto the Australian stock exchange. This accountant set about a valuation model based on a lawn mowing round of proprietary-based service industry, which gave an extremely twisted interpretation of the value of the company. The entrepreneur quite rightly turned the project down - based on the information he was given by the accountant he relied on. Unfortunately for him, this information was grossly flawed but he was not able to work this out for several years.

CHAPTER 7

Value from Licensing

"We think we have solved the mystery of creation.
Maybe we should patent the universe and charge
everyone royalties for their existence."

—Stephen Hawking

As I describe it, licensing means taking your idea, your "intellectual property", and giving someone with an established sales and distribution infrastructure the right to use it for a specific period, for an agreed fee. As the licensor, you set the terms by which you offer this right. In return, the licensee gives you an agreed payment at regular intervals, over the period. This is also known as a "royalty" cheque. These royalty payments, and sometimes upfront payment commonly referred to as a signing fee, (or one-time advances on future royalties) are the means by which you will generate income when you license your IP. These royalty fees can be between three to ten percent of the wholesale price of the product. If this appears low, it's because you make no investment in infrastructure, to meet the market demand. It is up to the licensee to provide all of this, as part of their commitment to building the value. In return, they get to name their price and brand the product as they see fit.

It becomes the responsibility of the manufacturer to advertise it, market it, manufacture it and invest in inventories and brand-building. Any potential licensee will have already an established network of retailers that they can plug your new product into. If you start a company on your own, not only do you need to raise money to start the company, you have to build a complete infrastructure, hire and train key personnel and build relationships with retailers so you can get your product into their stores.

This can be a problem since most retailers do not want to deal with one-product companies. Retailers want to deal with suppliers who can provide access to and inventories of hundreds of products they need to buy. Many retailers do not list one-product companies because they fear they run the risk of running low on stock or not being able to afford to fill a large order.

When you license your product, the manufacturer will have already established those important relationships with retailers, so all they have to do is plug your product in. It is far more cost-effective for a wholesaler and/or manufacturer to license a product or concept, so that they do not deal in the research and development aspects of new products, where they may not have any experience or know-how. As you build a relationship with a licensee, you may find that they are your first call for your next project, as well is them bringing new ideas for you to build projects specifically for them.

The most significant advantage of taking the licensing option is the reduction in capital required to build the business model. You can reach a cash flow positive position with as little as the cost of the provisional patenting and legal contracts. The operating costs are really only represented by ongoing patent protection and the cost to audit and/or enforce the licence agreement. If managed correctly, the patent-holder or licensor requires no staff to operate and would receive monthly or quarterly royalty payment to cover the cost of the licence. This is effectively like renting out the intellectual property for someone to build and sell products based on that design.

In most cases, patent-holders believe they can earn substantially more money by commercialising intellectual property themselves. The reality is that there will be people far more competent at the manufacture and distribution of similar intellectual property. You could best be employed researching and developing your next big idea, while receiving an income from other people who commercialise what you have now. They wear the risk and they put all the establishment costs into their project, so they should have the majority share of the sales revenue. The licensee may have ten or twenty years' experience in the industry of your invention and may have customers who would readily buy from them, based on their experience and reputation. That's why it can be an advantage to consider licensing and partnering up with a big established company.

Under the licensing business model, your licensee (the company which you license your idea to) invests their money and takes all the financial risk. If for some reason they do not perform, you get your invention back and you can then license it to someone else. There is effectively no financial risk when licensing a product, when compared to manufacturing and marketing a product in your own right.

Negotiating and monitoring a patent licence takes very little time resource from your daily activities. If you were to consider manufacturing in your own right, not only would you have to come up with the capital required to establish manufacture and distribute and sell the product, you have to put your time in to manage all that resource. The returns associated with this would very rarely justify 60 hour workweeks, which is a reality you would face doing this yourself.

If you start your own business and manufacture a product on your own, you are going to be locked into that business for a minimum number of years. For at least the first two years, all of your gross earnings would need to be spent on promotion, to build your brand and make your product more of a recognised requirement, within the retail systems. Under the licensing model, the licensee handles everything that's required to run the business, so you'll have the free time to go about building your next innovation.

It's very rare that a licensee would attempt to breach a contract with an independent inventor. There are a few reasons for this. One reason is that companies in general do not want the liability and financial damages they would incur if you can prove they breached the agreement. Given that most patent attorneys would see the patent litigation process as a sure thing, in most countries there would be patent attorneys prepared to represent patent-holders on a pro-bono basis. This places the licensee at a disadvantage, if they are facing a litigant who has no cost limits. With careful and diligent records, the patent-holders will always be holding the advantage. Having said this, any patent holders should make best efforts to not end up in court. The more litigation you participate in, the more of a reputation you get as a patent troll, which will minimise your earning capacity from your ideas later.

Managing Risk

Given that the patent application is going to be the biggest capital costs associated with the licence agreement, timing of a patent application will be critical. Before one rushes off to lodge even provisional patents, you can conduct some extensive secondary desktop research on the industry, as well as building a features and benefits matrix of all the competitors who play in this space.

Next, if you have a working prototype already, it pays to talk with prospective licensees, under a confidentiality or nondisclosure agreement. The purpose would be for them to get an idea what you have and what it can do for them, as well as to give you an idea of how much money they can make out of this, if you enter into a licence agreement at a later time.

This type of market research can be invaluable to the developers. It helps you to show how the end customers will use the product and what iterations or variants they will require to tap into other markets. Getting feedback like this from a select number of key players, will not only give you valuable feedback in your development phase, it also enables you to present your idea to prospective licensees, while still at the concept stage.

More than ten years ago, I worked with an individual who had developed an idea for the plumbing industry. The industry was dominated by less than twenty companies worldwide, so this product was ideal for a licensing opportunity. We did a cornerstone investment with a company in his city, as the principal of this firm was enthusiastic about the idea. When the time came to license the prototype, we put the feelers out to all twenty companies and we travelled to four countries around the world to present the idea to the principles of these companies. One of every three of these companies was in the USA. When we arrived at the offices by appointment in New Jersey, we were met by a team of lawyers who presented us with a watered down version of our nondisclosure agreement. This effectively stated that the company may be working in technology in this area, and if they were they would be excused from any obligation under disclosure or application of technology. I closed the meeting immediately and then spent nearly 30 minutes getting the inventor out of the room. His combination of anger and desperation played to this company's strategy, but as an un-emotive third-party, I was able to dampen down the emotion and walk him away.

We were walked out to the car park by one of the marketing people, who casually asked us what hotel we were staying at. Sure enough, that night we did get a phone call, inviting us to another meeting the following day at another premises, which happened to be a subsidiary of that company but, we were advised had less restrictions on their ability to sign nondisclosure agreements. Ultimately, the deal was done with the subsidiary company and the product is still in production today, although royalties have long since ceased.

Proof of Concept

As mentioned in earlier chapters, the best proof of concept will be a working prototype, which can prove the commercial model of making or saving time and/or money. In the past, developing a prototype can be a very expensive proposition. However, these days with 3-D printing, there are websites such as e-Lance or O-Desk, where you can get people to bid on producing a 3-D design file and then get others to produce a prototype for you, all within a fraction of the cost it used to be for plastic prototypes under solid modelling techniques. If you are technically-minded, it might be sensible to invest $2,000 in an entry-level 3D printer.

The most important part of your proof of concept is not the prototype – it is always the business model. You have to be able to convince senior management in your prospective licensing company, that if they were to add this concept to their product or service portfolio as it is now, they will be able to generate a significant amount of extra income without incurring much more cost.

It could be that their existing customers would also buy one of these and they may have 30 years of past clients they can approach, to whom they can offer this service. This leverage has other advantages for the licensee prospect, including reinvigorating old customer lists and defending customer's loyalty, who will not switch away to other suppliers if they need this new innovation. Sometimes the most effective way to prove this case is to prepare presentations and testimonials on video, from their clients.

Pitching your licensing opportunity to big companies, should be handled in the same way as pitching a trade sale, which we have handled in earlier chapters. The only core difference is that you should pitch your ideas to potential manufacturers, far earlier than you would to investors in a commercialisation process. It is important to understand the potential of your idea before you build a working prototype. This includes the value of future sales, which can give you a rudimentary estimate of licensing earning potential. Just as critical, is the industry research. This will tell you who the competitors would be in that marketplace, what their substitute products are and most importantly, what their pricing position is now and would be after you for your licensee introduce this concept.

Realise that no one can be as excited about your idea as you. If you needed help in pitching to the right potential licensees, I suggest you get very familiar with E-Lance and O-desk, to have someone prepare a list of all potential licensees in your region, then have someone prepare the pitch deck presentation and pitch sheet for you. In most cases, it's going to make more sense for you to pitch the product, because you understand that better than anyone and you can answer any questions that are going to be thrown back at you.

Royalty Rates

The most common royalty rate for consumer products is around 3-10% of the wholesale price. So if a product sells at a retail store for $20 and that store pays $ ten for it, then you would receive 30c to $1.00 for each unit that is sold. This rule of thumb will enable you to plan an estimate of the potential income for any project, before you even commit to the prototype. If you can determine by interview, the number of people who would buy this product in relation to the number of products sold in the market space, then you can calculate very quickly what your royalty value would be per region and what they can mean for you in potential income. If you speak with successful serial inventors, they will almost always work to a licensing model and will have validated their licensee needs before developing or patenting anything. In many cases, they would ask current licensees what other problems they (the inventor) might solve for them.

One of the key advantages of a royalty structure is that your license can be a geographic or demographic segmentation. In non-technical terms, this means that you can license for a particular region, with a sub licence for a particular industry category. It could be that your technology could be used across more than one industry category and therefore you can have more than one licence per region which can still be an exclusive licence agreement.

When there are particularly segmented demographic territories, it could be that you would decide to commercialise one yourself, but you do a licence agreement with a company in another segment, and take your royalty income as product at wholesale or less. This enables you to be in the distribution and marketing business, without having to tool up for manufacturing.

I had a recent case where a company wanted to do global licenses and had a geographic territory of China, which is one of the hardest markets in which to enforce a licence agreement. What we decided to do was to find a manufacturer who would license the product, but enter into an agreement to supply the product they produced (for my client's other international markets) at cost-plus-10%, in exchange for exclusive territory license for China itself. What this did was to hand the policing of the China licence to that company, and in return it gave us access to the production rights without any capital or tooling requirements. We were able to then sell other geographic territories with an offer of supply, through this Chinese company who would benefit from the economies of scale provided by a global production.

Doing a pre-production licensed calculation is a quick rule of thumb to determine which projects you should focus on, in royalty volume, rather than speed to market. Once you have determined the potential for each of the products, you can park most of them and only work on those that will give you a healthy market share through licensing and will not constitute a difficult development process.

Licensing Performance

Licensing agreements, just like any other performance-based agreements, will need to have accountabilities and measurements built in to the process. This can be onerous and difficult, particularly in countries where the language is not your own. The annual audit process is an option that can eat your royalty before you start, so you need to think of a better way to collect money and not have your licensee withhold some of your earnings.

The best way to achieve this is to set agreed targets of minimum and best potential, on an annual basis for the product. Once you have this, you can calculate your royalty and agree to settle the minimum sum as a quarterly payment, regardless of their performance. This effectively prevents you from enjoying an outrageously successful product for the first geographic territory, but does give you the benefit of not having to audit the process on an ongoing basis.

For subsequent territory agreements, you could base your forecasts on the actual performance of the first territory, which will give your far more accurate estimate of your earning capacity, whilst still locking in a fixed quarterly royalty amount.

Timing

Timing becomes a very important feature of a licence agreement. Any arrangement must have clear milestones for each stage of the development, manufacture and distribution process. Any delays in the production and distribution of your licence will affect your income, but not necessarily affect those of the licensee. In cases like this, we can revert back to the minimum and maximum earning capacity agreed to in the agreement and to clear the maximum as the payment required once the product milestones fall more than 60 days behind. This effectively forms a performance penalty for the licensor, ensuring the licensee will not defer or delay the path to market, without good reason.

There are always built-in performance clauses to the average licence agreement. You need to be aware that these are both-ways. If your design has a flaw, and it requires a recall based on the licensee's customers refusal to accept it, you may be liable for losses suffered, unless you have been explicit in your licence agreement. If on the other hand, the volume of sales does not meet a certain amount each quarter, regardless of the agreed fixed royalty payment; you need to have an exit from the agreement. Most agreements feature an accelerated model for annual growth and a hand-back clause for non-performance.

It could be that some companies would take a licence on your product, in order to shelve the idea as it may cannibalise their existing product lines. That will reduce your market earning potential and you need to exit that agreement if you have a reason under the performance clause. In the event that somebody does shelve your product because it threatens their market share, you can expect that the competitors would be more than willing to take your licence agreement once you have extricated yourself from the current agreement.

Experienced licensing lawyers would very rarely want to include measurements of margins, agreed retail pricing, and other metrics. Having these metrics makes the agreements far less workable in enforcement, but not having them can lead to companies using price gouging to maximise their margins, at the expense of your volume of product. While this can be done in a protected market with a good solid patent, it does not work to your advantage. Although I do not advocate dictating retail pricing and other owner's criteria on any potential licensee, I would place a clause for annual review in any agreement, for working an economic model of price and volume that could advantage both parties.

It is important to recognise that although a licence may be in place, the intellectual property is still owned by the licensor. If, for any reason, the licensee experienced trading difficulties or bankruptcy, the title should not end up in their asset register for sale by their appointed receivers. Most licence agreements have hand-back clauses based on liquidity and liabilities, which shift the licensing rights back to the licensor.

Savings and Shortcuts

Although I always advocate the use of legal and accounting professionals in the development of business systems structures contracts etc., I am always one for sourcing a good template for licensing and other set boilerplate type operations. You can generally do this online, by buying a boilerplate licensing agreement from a number of legal libraries, and then using online services such as e-lance to modify this to what your requirements would be.

When you are ready to take this to a licensing attorney, you will have more than 90% of the work done for that particular jurisdiction, which will reduce your number of hours the attorney will take to build you the final document. By now, if you are committed to working this model, you will have already allocated that 90% saving in legal fees, to independent market validation.

In most cases, if you produce a draft agreement before the other party, your draft agreement is more likely to be used as the foundation of the final agreement. This framing can be very important advantage for you, by eliminating many far-fetched options that potentially could considered by an inexperience licensee.

One of the core temptations by many innovators is to load a substantial commencement fee or signing fee into the licence agreement. I generally prefer not to have a commencement fee loaded onto the front of any licence agreement, because I see this as risk for most of the licensees. Most innovators try to use a sign-on fee as their method of recouping their development costs and lost hours, with the hope that their royalty payments will then be pure profit. I have never yet met an innovation that could justify that claim. While you may see that as a prudent business position, be mindful that the prospective licensee may view this as pure greed.

Revoking Licenses

Getting out of a poor-performing licence agreement can be onerous, unless you've built in an agreed performance criteria and a hand-back sequence, to the licence agreement. You will also need a clear exit process, which is automatically activated, unless you agree otherwise, in writing. This will eliminate any delays associated with legal jousting, as the licensee will struggle to retain the licence and you will want a quick exit so you can appoint someone better.

Most boilerplate licence agreements should have the sort of revocation clauses built in. It's up to you to modify this and add in features you require, before you hand it to the licensing attorney, who will then massage it into a professional and enforceable agreement. I recommend you change out the word revoke or revocation, to "hand-back."

Like revocation, renewal clauses need to be built into the initial agreement. It could be that the licence agreement may automatically renew, based on all milestones being met throughout the first period. It is not always an advantage to have an ongoing licensing agreement, just in case for any reason, the spirit of agreement does not prevail.

Future Products

In most cases, it is essential to have a clause in the agreement that excludes future products from any licensing agreement. You must have a clear option to licence to anyone, for any future products. There will be a demographic restriction on products you produce to be distributed under a separate licence agreement. The concern is more about the right you have to build a different widget and have another company in another territory in another market, produce and distribute that for you.

CHAPTER 8

Crowdfunding

"Crowdfunding is not new. Most people do not know that the Statue of Liberty was crowdfunded....."

—Erica Labovisz

Why Crowdfunding?

As integral as capital-raising is to the intellectual property commercialisation industry, there is an expectation that crowdfunding will become integral to the capital raising industry, within the next five years. Crowdfunding first exploded onto our scene in around 2012 and has progressed rapidly, despite the regulatory opposition by most jurisdictions. Many of the tax offices and stock exchange platforms and their respective regulators consider this a major threat to their highly-regulated capital raising industry. Most likely this is because they are going to miss out on their commission revenues as a result of the virtual nature of an online capital platform. There is the very real risk from some parties presenting fraudulent opportunities through these platforms.

Having said this, most of the platforms these days have a highly automated due diligence system which is able to capture most of the fraudulent projects before they go live. As a result, more and more amateur investors are starting to look at this platform as way of jumping into ground floor opportunities, with only pennies at risk.

Some countries do not allow investors to buy equity in projects, but they will allow the project owners to pre-sell their intended products. This enables the inventors to put up first run products at a discount, in exchange for the pre-payment, to be used for tooling and first bill of materials. These platforms have the potential to shift venture capital into the hands of the public, without the leverage of financial muscle that some venture capitalists may use to extract exorbitant percentages of equity in promising projects.

Do Your Research

If you wish to attract a modest capital base, you have a direct application for the funds and your project will show up very well on video, then a crowd-funding platform might be perfect for helping you to raise the capital you need quickly.

Before you make a decision to jump on one of these platforms, you need to do your homework to ensure that this would be the better option for you than, for instance, a Cornerstone Investor. Although a crowdfunding platform does not bring in any knowledge or experience on the product or industry, it can easily offset this with little or no demands for control of your project, should you fall behind on milestones.

The first part of research is to select the most appropriate platform to use. I would personally recommend Quirky, as it is designed specifically for innovation. This means you have to stand out from the crowd in a very special way, so you really need to have something that is visually appealing, to ensure you can capture people's attention long enough for your video to explain to them why they cannot afford not to invest in your project.

One of the more critical advantages you have from reading this book is to present your business model before your product, which will set you above the crowd very quickly.

It is important to understand that most crowdfunding platforms cater to different service segments. Non-profit platforms do not service business and innovation platforms do not service music events or movies etc. As the industry gains acceptance, business owners are using different crowdfunding sites than musicians. In turn, musicians are using different sites from causes and charities. Below is a list of my top 5 crowdfunding sites for 2015, which provide a platform which is suitable for intellectual property. These may not be the largest crowdfunding platforms, but they are the top 6 in the category that would best cater for intellectual property in the world today.

Some of the top sites in the world today (starting with Kickstarter and Indiegogo), have a large cause-based and/or donation-based crowdfunding base. They do amazing work with fixing small financial problems for charities and groups in need. For this analysis, I am focused on just crowdfunding platforms which can and have catered for funding the commercialisation of innovation. These include:

Quirky

Quirky seems to dominate in the innovation space and has lots of great ideas with some being presented very professionally. It can be seen as a place to collaborate and crowdfund for donation-based funding with a community of other like-minded folks. Quirky digs deeper into the commercialisation process and will still attract interest at very early (pre-value) stages. It has grown to become a community for inventors and ideas are shared rapidly through their viral networks.

Crowdfunder

Crowdfunder is the platform for raising investment (not rewards), and has one of the largest and fastest growing networks of investors. It was recently featured on Fox News as the new avenue for crowdfunding after the story about a $2 billion exit of a crowdfunded company was reported. After getting rewards-based funding on Kickstarter or Indiegogo, companies are often giving the crowd the opportunity to invest at Crowdfunder to raise more formal capital placement rounds. Crowdfunder offers equity crowdfunding from sophisticated investors, angels and VCs, and was a leading participant in the JOBS Act legislation in the USA.

Appbackr

If you want to build the next new mobile app and are seeking donation-based funding to get things off the ground or growing, and then check out AppBackr and their niche community for mobile app development. More of these platforms are springing up every day, but AppBackr is the most established in that space.

AngelList

If you are a tech Startup with a shiny lead investor already signed on, or looking for Silicon Valley momentum, then there are angels and institutions finding investments through AngelList. For a long while AngelList didn't say that they did crowdfunding, which makes sense as they have catered to the investment establishment of VCs in tech start-ups, but now they're getting into the game. The accredited investors and institutions on AngelList have been funding a growing number of top tech Startup deals.

Indiegogo

Having been in the industry since its inception, this platform has the most established following and some of the widest variety of opportunities. There is a solid audience for innovation, but there are also a lot of crazy ideas competing for a share of their money. If you stand out on this platform, you could literally attract millions of investors, or pledges for forward orders.

These crowdfunding sites cover most of the variations of innovation funding requirements you might have. It is always good to cruise these sites and look at the way people present their innovation and how this can be received by investors. Certainly, if you allocated more of your limited budget to a fun video on your innovation, you would receive greater interest than if you invested in an independent accountant's report, if crowdfunding was your choice of capital sourcing. This suggests that most investors on these sites make emotional decisions behind the anonymity of the Internet so presenting an emotional offer will always generate more interest.

Once you have selected the platform, you should work your way through at least 100 presentations previously completed on this platform, starting with the ones that have attracted a substantial amount of funding. Some presentations that have well exceeded their initial targets may have already been closed, but will be archived for reference. You need to learn just what these people did, that set their campaigns apart from others, given that it's less to do with the product or gadget, and more to do with the people and the way the project is presented. The first decision criteria will always be emotional, when you take the face-to-face meeting out of the equation.

The next stage is to select the most achievable amount that you are going to raise. Be mindful that if you set a target of $10,000 and only raised $9,900, your project does not succeed and you are not awarded anything. However, if you set a target of $5,000 and you raise $9,900 your project will gain far more attention and could attract a crowd, some of whom are there to see how you will do, while others will eventually decide they want to be part of it.

The next key point is how you are going to present the product. It is important that you engage an expert in this area to prepare a short two-minute video that motivates the viewer to click through and read material about your project. This has to be the most compelling call to action that you can muster, in a two-minute video. do not for one moment believe that your product will create that compelling call to action. You need an advertising expert plan and script this video and I can tell you there are thousands of highly qualified advertising creative directors sitting in E-lance and O-desk, who can prepare a killer presentation for a fraction of what an advertising agency would charge you.

Your Target Audience

Your next task is to clearly define your target audience. You must assume that any opportunity that is posted on the site may get looked at (in title only) by more than one million people a day, but perhaps one tenth of one percent would click through and watch what you've put up there as a description of your project presentation. You need to define who you are intending to appeal to, to urge that click through. The more narrowly-defined the target audience is, the more your message will appeal and inspire them to get involved. Most people in service based businesses fail to understand the benefits of targeting a very specific niche. The Internet is a wonderful method of segmenting markets, because when we surf the Internet, we have an average attention span of about one and a half seconds on any particular landing page, before we move on or make a decision to stay. Sometimes that decision is to look further down the page, but we only do this if the message is compelling enough and so defined that it appeals to the "me" inside of all of us.

To get the best understanding of our next target audience, you should start with dissecting the common traits of your top ten clients. If you have a consumer product, you have to work out where these people go, what they do, who they do it with, and how they would normally find your product, on which shelves and in which store. You may even be as specific as defining the types of clothes they wear, because any good advertising creator may use this information to reflect how he dresses the key person in your video.

Planning the Event

At first glance, crowdfunding websites can appear to be similar to eBay and other auction sites. There is a temptation for innovators to complete a listing application, post their product opportunity up on the platform and then wait a week to see how many millions of people subscribe to it. It would come as no surprise that very few of them would have even got looked at, if they do not have the compelling offer that makes them stand out from all the others.

Proper planning for an event like this will require at least two weeks but more likely five to six weeks, in the lead up to the listing, and then at least two weeks for the actual listing period itself. People who start a campaign without an audience will struggle to get any sort of traction. Ignoring how good the product offer is, if you do not stand out and nurture any audience, nobody is likely to see you above the noise of a million other opportunities.

The campaign would need to be broken down to three different sections of off-line, online and on platform, as three separate audiences - requiring three separate messages. These are further segmented into (a) before the start of campaign, (b) during the campaign and finally (c) after the campaign. Each of the segments has separate messages and may require a mixed delivery system, to maintain the interest and to pique the curiosity in every potential audience.

Campaign effectiveness and process will differ over time, so what works today may not work in a year's time and will most certainly not work in five years' time, if crowdfunding as a practice is still around. It is therefore not appropriate to detail a rollout plan for crowdfunding, as we may do for clients in our management consulting practice. The following are just highlights of what needs to be done in order to ensure you maximise the impact for your crowdfunding campaign. Treat this as a checklist of things that should be addressed, rather than an absolute todo list.

Corral the Hungry Herd

Once you have your target audience defined so precisely, you will now experience best to find them. Some of your more promising prospects may not be on the crowdfunding platform that you have nominated, some may not be on the Internet at all, but it's up to you to draw those onto the Internet and then into that crowdfunding platform, to play the part you want them to play in your program.

The first part of this is to draw up your list of prospective buyers, who fit the audience profile you are targeting. It could be that you have a product that is ideal for practice managers for accounting or legal firms and although most of these people would be online at work, some may not even have the Internet at home. A LinkedIn campaign will be able to reach these prospects.

For this purpose, your all-encompassing early campaign should include purchased contact lists from independent list brokers, and other parties, including successful online entrepreneurs who rent their lists for a proportion of earnings under collaborative agreements.

You should be able to get statistics of conversion for similar products that have listed on the platform you have selected. This may include the number of people who looked at the opportunities, the ratio of those who clicked through to examine the opportunity more closely and finally, the number who made an application. This would be the same for direct mail response and for third-party lists you may incorporate into your campaign.

If you cannot walk into a crowdfunding promotion with at least 1000 followers before you hit that platform, you are not going to create any traction within the period you may have set for yourself. If you are able to launch your opportunity and have immediate traction with some of the prospective investors you brought with you, the audience on the platform will take notice of the activity and pay closer attention to your opportunity. Given that close to 100% of the audience on the platform are potential buyers, you are more likely to get your video viewed and your opportunity scrutinised, if everybody in there sees you as the next hot thing.

Social Media

One of the easiest ways to develop your list before the campaign would be specifically targeting lists of your prospective buyers, on social media platforms such as LinkedIn, Facebook, and other specialty sites. A promotional page with advertising on Facebook or a join-me campaign on LinkedIn, can be set and automated so that the only thing you are required to do is monitor the statistics to determine what the most effective set of that advertisements to run over the four week period before listing the product are.

It could be that you know that all the prospective buyers in your target demographic may own or seek to purchase an aligned product.

An example may be a new fishing rod that you have developed and you can be certain that a high percentage of people in boating groups or looking at camping stores would have a fair degree of overlap to fishing rods. As a result, you can join boating groups and fishing groups on some of the social media platforms, in order to make mention of the upcoming opportunity of the new technology. This creates enough buzz that people will switch across and look at the campaign when it comes online, if they've seen four or five messages over the weeks before.

There are many savvy journalists with weblogs (known as blogs) who would be prepared to write about some upcoming technology, if the price was right. Some of these have audiences in excess of one million people, scattered all around the world with that one pastime, hobby or activity in common.

There are many advantages for going "old school" and engaging a publicist or public relations firm, to seek interest in mainstream media. Timing for this sort of activity is very important, as memory for some particular media is far shorter than for others.

Once you have generated some interest before the event, you still need somewhere to send these people, to maintain their interest and to feed them additional information efficiently. The best way to achieve this is to have a series of landing pages, which can be described as very small websites which are specific to each of the different profiles that you've prepared, for your new product. The objective of these landing pages is to give people a sample of what could be in store for them and then encourage them to ask for more information by clicking on a request and completing a form which includes their contact email address.

Once you have the contact email address, you start to build your list which we would have referred to in many of the chapters as your hungry herd. It is important to keep the hungry herd fed and for this reason, you need to have updated information in a timely manner to keep the message on track and to keep the interest through maintenance of some suspense. You may have several different landing pages for the one campaign, each to different target audiences, offering different messages and outcomes, for the same product. What works for some, will not work for others and so having different messages will still bring the same crowd to the same party but for different reasons.

Once you have your list starting to build - well before your launch campaign is underway - you will still need to nurture the list with teasers and other seeds of information. This would include regular automated email responders, which could be prepared well in advance. You may sequence these for every four days and target specific groups for specific messages at specific times of the day, in their own time zones. This can be a set-and-forget exercise, if it is done by professionals. Again, you can find these on some of the labour hire websites such as E-Lance and Upworks.

Some of the interactions you have with these prospects might be used to gather further information from those that are more active and keen. This might be through offering one or two questions each week, buried inside the personal email. Being able to process the responses can take time, but the advantages of narrowing the target demographic will make the advertising far more valuable when you promote the crowdfunding campaign.

Your Compelling Exchange

Well before any crowdfunding campaign, most of the promoters would decide the amount of capital they would like to raise and what considerations they will offer to the investors or buyers. What will they get for jumping in early? How is this offer different from waiting until this product is available from their local store?

If you treat your crowdfunding campaign as a cold transaction, you will not get the sort of urgent activity you require, to make these campaigns fly like an auction. You have to carefully craft your offer including what privileges you are going to provide to every member who signs into the program and participates in your opportunity. You may offer a discount on the retail price, and extended warranty or a special investors-only limited edition copy of whatever it is you are going to produce. Whatever your special offer is, it must be compelling enough to generate an instant call to action, for all viewers, as they watch your video and read your offer.

You have to then set a realistic time for the amount of sales or capital that you require. In most platforms, if the minimum target is not met, the transaction does not occur and all funds are returned to the bidders. I suggest you start with a low platform but accept oversubscriptions. I'm sure if you were to set out to raise $100,000 and you received offers for $2 million, you would not consider that a disappointment and you may adjust your ambitions from immediate sales to "world domination", or something similar.

The Video

The most critical part of the crowdfunding campaign, is the video. To understand how well the video can work for your promotion, you need to set aside a good four or five hours and log into as many of the crowdfunding platforms you can, just to watch the videos and listen to how people present. This does not mean that you (as the intellectual property owner) should have to do the presentation, but it will give you a feel for how successful promoters offer their products and what sets them above others. The underlying theme for any good crowdfunding video is that it must be short, simple, and concise. It's great to have humour and a compelling human interest story built into this, but never more than two and a half minutes of lively and compelling entertainment. Your secondary objective is to have the viewer send a link (of this video) to their friends, thereby creating a viral response.

When crafting the script for your video, it would pay to include two endings for the video, featuring one to be run before the campaign begins and the other for during the campaign itself. This saves your update from looking like a tacked on adjustment done after the project is underway. A third, always a good option, being an ending that says "We have reached our primary target and we are pushing on to achieve our secondary goal of this much".

If your team is critical to the project, you might wish to present them in summary, just as you would introduce them to friends of yours. You do not need to go into their history or even their roles within the project. That can all be handled on one of your landing pages or the official website. The video is to capture the attention of viewers and make them stop to read all the other information you want to give them.

The format you need is not a cut and dried suggestion. It will change over time, so there would be no point in putting an already dated format in this book. Instead, I believe you will get the best format from the more successful videos you watch in that four hours of platform research you need to complete. Get your business partners and some popcorn and settle back for some (mostly) less than riveting entertainment.

Above all, keep your video informal and personable, because the people watching this must engage with you on an emotional level to be able to accept what you are telling them and take action on this within a two-minute interval. This means that showing a PowerPoint presentation across the screen is most definitely not your first choice of entertainment. It does not matter how good your project is in comparison to any other similar project or any other project on that platform. You must remember people are on the Internet to be entertained and will move on in seconds if they find your message to be less than entertaining.

After the Campaign

For this exercise, let's assume that your crowdfunding platform campaign was an outstanding success and you are about to kick the project into top gear. You must remember that you now have stakeholders that you didn't have last week, and stakeholders need information on a regular basis in order to maintain the level of satisfaction they have with the decision that they made. This reinforcement and reassurance of their decision serves to eliminate any future doubt they might have and keep them engaged with your project as an ambassador in the future. The ultimate objective is to have them share their involvement with their friends and contacts.

Providing an online blog of activities or a distributed newsletter might help to keep these people informed and may even serve to empower them to bring you more buyers and/or investors in the future. Sometimes you may struggle to find things to discuss with investors or stakeholders but I would suggest that just the daily routines or daily activities and the allocation of the budgets may appear to be mundane, but would prove newsworthy to investors. When this involves keep investors up to date with their investment it will always provide a measure of satisfaction that they put their money in good hands. This level of accountability can prevent a public relations disaster at a later time, if any of your stakeholders become disenchanted with your progress, for no reason other than they have not been kept up to date.

Always be mindful that all of these stakeholders, including those that engaged with you before the event but didn't invest, are still members of the hungry herd and can buy more products when your production goes into full swing. They engaged with you for a reason and it's up to you to identify what that reason is, in order to motivate them to make a purchase decision. The longer you keep the conversation going with them, by updating on the progress of their project, the more opportunity you will have to market your products or services to them in the long-term.

CHAPTER 9

Scaling Up

"The critical ingredient is getting off your butt and doing something. It's as simple as that. A lot of people have ideas, but there are few who decide to do something about them now. Not tomorrow. Not next week. But today. The true entrepreneur is a doer, not a dreamer."

—Nolan Bushnell, Entrepreneur

Target Market Focus

So you have completed the research and development, raised the capital to commercialise, set up the company, and now you are ready to test the market. How much is this innovation going to earn you as a shareholder? Firstly, you need to define your target market. You need to know who will buy this and why.

In most cases innovators want to keep their target market wide, so they can offer their product to significantly more potential buyers. If your target audience were narrowed down to a group who could readily understand what the product did and how it could solve their specific problem, you would have more sales and quicker decisions.

One of the best reasons for narrowing your target audience for a particular product or service, is that the narrower the market is defined, the fewer competitors you will have who will identify readily with the buyers in that space.

In our innovation on Formula1 for Business, my business partner and I developed our twenty years of intellectual property into online coaching and consulting tools. This gave small to medium businesses a risk-free and affordable way to double their net earnings in a ten-month period, using specialist tools and checklists. We had our own test and measurement tools and project status review processes, and we delivered our weekly Q&A meetings by Internet. This way we could provide a service to many - with no geographic boundaries.

As consultants, we value coaching and consulting. After one of our advisers (a leading business coach in the world today) unpacked our program and examined it, he suggested we adjust our focus to 'narrow and deep' in order to find a specialist niche market. We agonised over this for some months, but came to realise it would enable us to focus on the types of businesses we enjoyed helping the most. These happened to be service-based businesses - in particular engineering.

In the end we set up websites and marketing collateral around helping engineering practices overcome their most significant issues. We were well aware of these from our dealings with engineers over the past 25 years. I approached this problem from the commercial perspective of the intellectual property we had developed through prior experience, Simon, on the other hand, approached it from a capital raising and management perspective, consulting to intermediate and larger engineering firms across Australia.

We found it easy to tap into these wants, needs and aspirations, and quickly modified our intellectual property to reflect these needs as they related to engineering firms. We had effectively eliminated almost 90 percent of our demographic market by insisting on engineering firms alone. However, with the birth of the Internet, we could extend our geographic reach to include other cities, other states, other countries and, eventually, other continents. It took us nearly five months to change the business to a specific niche, but with that niche we generate considerably more than we did as a generalist business to a wider audience.

The idea is to own the market niche you are targeting, and the best way to achieve this is to know that (a) there is not a competitor dominant in that niche already and (b) your solutions to their specific problems are both recognisable and effective. It helps if these solutions are cost competitive with the solutions they are using at present.

Your Competition

A significant issue overlooked by developers and innovators is their assessment of competitors. If there is one area of your project that investors are going to look at more carefully than the product itself, it will be who you perceive as your competition, as compared to who they perceive your competition to be. Reasons for this have been documented as far back as the 1920s, when investors focused on railway stocks and tramway stocks as the most sustainable long-term investment. They could not even begin to imagine how any 300-ton vessel could travel at over 500 miles an hour at 30,000 feet in the air, transferring up to 600 people from one country to the next, in a matter of hours. When you prepare your list of competitors, look at other issues such as how many alternative options there are. If I were to start a taxi service from the outer suburbs of the city, I would look at what other modes of transport were available to people in those outer suburbs, including buses, trains and their own vehicles. Then I would look at the telecommuting as an alternative where people do not physically go to work but use their computers from home. To assess your competition with any seriousness, you need to know who your main competitors are, why they are competing against you at the moment and whether (and by how much) their segment of your market is growing or contracting.

After you have allocated a value to each of these market players, you need to ask how committed they are to that industry. You also need to be able to justify to investors and others why you hold these views. Third-party validation is the most critical component of your marketing plan and must be completed before you start to estimate how many services or products you intend to deliver to this market niche with your innovation.

Next, you need to assess how you will access this market and what resources you will use to present your solution to the decision makers who have the need. Although it happens less often these days, I used to see innovators presenting their marketing assessment in these terms: "…this market segment is worth so many billions of dollars and we are targeting only one percent of the segment. So the value of our company should be this."

The marketing plan is not just an internal document. Independent validation is most important for external third parties. These could be the investors, or any strategic alliance partners you intend to recruit. The marketing plan could also be a recruitment tool for sales or marketing personnel. Also, during your commercialisation process it could provide justification to the board or technical advisory panel when you are requesting an allocation of interim funding. It is important to get this right so you can access the resources you need when you need them, and not have the accuracy or authenticity of your estimates questioned. Having independent, third-party research as a backup should alleviate most critical questions.

When looking at your competition, you will benefit if you have researched what the market is prepared to pay for each of the current solutions. This is because most buyers switch brands or services on the basis of price, and if you can show that your option is cheaper at the time of purchase and in the long term, you are well on the way to secure the confidence of investors, partners and fresh recruits for your team.

Path to Cash Flow

When kicking off the campaign, you must allocate your marketing budget wisely. A fatal error that innovators can make in this regard is to try to market their product to several different niches and try to solve problems with unrelated industries, sometimes just to prove their product can do it.

Every marketing planner's aim should be to allocate all the marketing resources to what we call the "low hanging fruit." These sales are the easiest to close and they can provide an early cash flow and a sustainable growth pattern in the short term. If you build one niche at a time and receive a cash flow from each one, you can increase your reserves until you are ready to launch a full-scale marketing campaign.

Forming Strategic Alliance Partnerships

At times, your best market opportunities may be through retail stores. If you are a single product company with no track record, you have little chance of being listed as a supplier and receiving a stock-keeping unit (SKU) number for your product from the retailer or wholesaler. If you have just one significant primary market, you will need a strategy to get your product into a retail environment. That may mean partnering with an existing supplier to that retailer. In this way you will get a supplier number and a stock-keeping unit number for your product so it can be listed on shelves in their stores.

Sometimes your ideal partner may be a wholesaler who supplies to that particular retail chain. In most cases, however, it will be a small supplier to that store group who already has their product range listed. The most appropriate way to identify any potential suppliers to the retailer you are targeting, is to first check the shelves in the section where you believe your product should be listed and to ask the store's sales manager for referrals to suitable suppliers. They may know the key personnel first-hand and be able to open up doors for you.

Another method to gain effective strategic alliance partnerships is to look upstream and downstream at the goods or services that relate to your product. The key here is to identify what the end buyers will purchase just before and just after your product or service, in order to enhance their experience or achieve their goal.

If your product or service is of small value, you could even look for high-value suppliers who might want to add it as a premium to enhance their own sales. For example, if you have a fishing innovation such as a rod or rod holder, this could be an attractive product to a retailer of pleasure boats. Many retailers of high-end products like to provide extra incentives to prospective customers, and your product could be just what they are looking for.

Group Networking

The key to most small businesses is managing growth. In the majority of cases this means looking after your expenditure closely. In a fast-moving world with a high-growth business, you need to find cost-effective marketing tools which will put your business ahead of the pack, but not cost you major capital expenditure up front (such as advertising campaigns). Networking is the first marketing tool you should use to grow your business strategically. This involves your time but little capital outlay.

A few years ago, as part of a pre-start assessment for a Formula One client, we unpacked his current client list to identify where his major customers came from and which of his methods for developing the business had been most effective.

This client had an engineering consultancy business in a technology field and he was dedicating a lot of time to attending business functions and industry networking events. Two of his largest clients actually had children at the school his kids attended.

After we analysed how these opportunities became sales, it was clear that this man was providing quality information to these prospects very promptly, in order to maintain a good reputation with other parents at the school. In doing so, without realising it, he had become more effective at networking with these parents from his kids' school, than he was at networking at the industry events.

A key reason was that he was clearly not always following up on the opportunities he obtained at regular events, in a timely manner. He also wanted to preserve his reputation at the school.

His first thought was to start vigorously networking at the school, until we pointed out to him that he had been more effective because of the quality and timeliness of his follow-up. We showed him how if he had handled all the opportunities the same way, he would have done far better with the contacts he had made at the other functions. This might sound like common sense in hindsight, but at the time it was an eye-opener for him. We proved this with our subsequent Formula One program for his company.

We need to point out however that networking can be a double-edged sword. Although it costs little in capital expenditure, it takes far more of your time (a more valuable resource) to be effective. You can of course outsource some of the networking to your senior colleagues, but it does lose its potency if not done by the principal.

The most common reason for failure in networking programs is the follow-through. Lots of people go to events and hand out cards, but they (a) do not hand them to the right people, (b) they do not identify why these companies should be talking to them and finally, (c) when they get back to their office, they put all the cards in a drawer and wait for those people to call. Getting a contact from somebody who want what you have, will not generally happen.

We understand why it will not happen, but we still make this mistake. Which is why networking is so important to the few committed company principals who understand that if they do follow-through with new opportunities and people, they can generate a substantial amount of income with little competition. If you take away just one item from this article, it should be that *the money is in the follow-up and not in the networking event itself.*

There are several different types of networking but we will deal only with the more common and more effective ones. There are two distinct categories: face-to-face and online.

Face-to-face networking is the act of meeting people in a business or social environment, where you agree to exchange contact information for future contact. This process will get your project in front of far fewer people, but they are generally more qualified in the process as you chat to them for ten to twenty minutes at a business event.

Somewhat less effective is online networking, using such tools as LinkedIn, Facebook, Twitter, Instagram and others. This lessened effectiveness is offset by the sheer numbers you can network with, without any effort. From the perspective of professional business development, we have found LinkedIn to be pure gold when used properly to grow engineering firms.

Online networking will enable you to publicise your activities to thousands (or even millions) of new people, in other demographic or geographic territories, who can connect with what you are seeking and/or offering. It has revolutionised networking itself.

When we think of face-to-face networking, we generally think of cheesy salesmen with a glass of wine in one hand and a fistful of business cards in the other, all trying to talk over each other about themselves and their businesses.

These people would do better to wear a sandwich board and saying nothing, because their conversation generally has nothing to do with business and everything to do with them. If you become trapped talking to one of these people, they generally do not experience to drive or end a conversation, and will generally kill your prospects for the evening. Excuse yourself early and move on.

The biggest mistake business owners make when they are planning a networking campaign is to join associations that they, and by definition their competitors, could or should be members in.

If you have an automotive repair business, you will be less effective if you join an automotive repair association than you would be if you joined a business council, chamber of commerce, institute of management or other services, or other professional service groups. Networking within your industry is less effective because most of the people in those meetings either compete against you or have relationships with companies that compete against you.

Networking within formal events, such as in a stand-up cocktail party event, with ample opportunity to mix with people you do not know, can be an effective way to gather leads for immediate or later qualification. In most cases, people stick with a friendly group and can miss out on meeting a greater number of the others present.

There are two critical elements of a networking conversation: how to begin it without reference to yourself, and how to end it politely so you can move on to the next group and continue your work.

Some general rules that can boost your effectiveness in this sort of context:

1. Wear your name tag on the right lapel of your jacket. This makes it easy for people to read when you are extending your hand to shake theirs.

2. If you are to carry a plate of something and a glass, it always helps if you balance your glass on your plate and put your finger food around it. Then you can carry this with your left hand and shake hands with your right.

3. Don't push your business card in someone's face within the first few seconds of meeting them. Get to know a little bit about them and their business. Look and sound as if you have thought about it and decided you might be able to help their business, or they might be able to help you.

4. After introductions of names (and there may be several people in the cluster to whom you are being introduced) you should ask a question which does not immediately presume that they are in business or have a business that they are there to promote. My favourite opening line is "So what is your story?" I get a huge variety of responses to this, from things that went wrong that day to new additions to families and sporting events people have just won.

5. Get people talking about themselves; Our favourite subject is always us and if you can get people to talk about themselves, they will feel more comfortable in your company. This does not mean you ask them to tell you about their company to let them ramble on for thirty minutes about how they change paper in the photocopier or prepare a set of accounts for other small businesses. Small, inquisitive questions can move the conversation along different paths, making it interesting to other people in the group as well as helping you gather information on what they do and what they are most proud of.

6. The act of exchanging cards is not about handing someone your card and asking for theirs. The conversation should be a conclusion such as, "From what you tell me, I think we have a lot in common and it might be worthwhile for us to get together sometime and chat about opportunities we

might share. Here's my card, Bob. Do you have a card for me please?"

7. Once you have chatted to the people within that cluster who are of interest to you, you need to excuse yourself and graze in another paddock. A strategic exit could be framed like this: "....well I can see you guys have a lot to talk about, and I will be looking forward to catching up with some of you, so I will leave you to chat."

8. In a one-hour cocktail party, you might effectively cover up to four groups. This might give you up to twenty business cards, if you accept cards only from people who could be qualified as meaningful. The key is to keep moving and always be ready to graze in the next paddock.

9. It is good etiquette to look for individuals or parties of two who are looking or feeling lost and invite them into another group, by introduction. For an individual standing on their own, you might ascertain that they are not waiting for anyone and say, "I was going to join this group of people over here. Would you like to come with me?" Bringing someone into a larger group is the perfect icebreaker for them and it will give you an opportunity to speed-date the members of the new cluster, using the above techniques.

Another effective networking process is the organised breakfast, where similar-sized businesses gather in an organised structure and present one after another with a one-minute talk on who they are and what they sell. If you have fifty people in a room, this will burn up an hour and a half of your time with no guarantee that you will learn who you should be doing business with, within that group.

To make it pay, you need to take one of every member's cards and make notes on their company if they have something of interest. When you exit the event, you may have fifty cards, but ten to fifteen people with whom you could make an appointment later for a coffee or lunch. do not try to do business on the spot. Your objective is not a sale – it's a qualified appointment.

The most significant method of face-to-face networking is collaboration. This requires careful research before any conversation and it results in a meeting or conversation which is directed to a specific outcome by both parties. To make collaboration effective, you need to understand a great deal about the prospective collaborator, how they market their products or services, their target markets and how the same markets could utilise your services without being a threat to the collaborator.

This method is used by venture capital companies and large corporate consultants, to leverage fast and effective growth programs, for greenfields projects or commercialisation projects. If you were a distributor of alloy wheels for vehicles, an ideal collaborator would be a tyre shop, a wheel alignment place or a workshop which specialises in customising vehicles.

If they do not have products that compete against you, they would make excellent collaboration partners. We look for the market reciprocations (how we can offer their products to our clients and our products to theirs) before we then start a collaboration on winning new business with a complete package.

The online networking process is more complex, and requires an actual strategy. Like the offline variant, networking online is about identifying the issues of the people you are connecting with and offering them unmistakable value.

The beauty of online networking is the possibilities it brings for specifically targeting individuals with the right background and connection to further your own business prospects. In most cases, just getting past a "gatekeeper" on the phone at reception would render some key contacts unreachable. Now all this has changed

By way of example, in 2014 our Formula1 for Business program had a real estate client (sometimes we help those outside of the engineering space, who impress us) that was establishing an operation in Australia. This company was already in Singapore and operated across more than twenty countries. They had little experience with social media and wanted to establish their online marketing portal for real estate agents quickly. We chose LinkedIn to establish a foothold in this market.

At the beginning the local agency principal had five other members who had linked to his LinkedIn account. Three of these were family and another was me. We rewrote the LinkedIn profile, added some e-book articles and a short video, and set a virtual assistant (at four dollars an hour, working from the Philippines) to link him to more than 1,000 people in the real estate industry in Australia.

She had over 1,400 members linked to him in ten days, with over 280 of these downloading his article and responding to his automated message thanking them for linking to him.

This was achieved in less than ten days and set the company up for a fast start in Australia. This company already had an International profile and were established in more than twenty countries, but they still started this from scratch. They had more than a thousand qualified industry people to talk to in ten days, and more than 280 of these were enquiring about doing business with them.

Using social media for networking can be a powerful tool but it can also be very wasteful if not managed by experts. Most of us who commissioned a website more than, say, seven years ago spent tens or hundreds of thousands of dollars to get a nice-looking site up and running, but even today it might still not generate income.

Websites can be a waste of money if they do not have specific tools to direct people to make decisions. It is the same for most social media.

Individual Networking

After every group networking event, you should have a number of business cards of the people you met at the event. I generally sort these into four categories, being (1) people I could do business with, (2) people who could refer prospective clients to me, (3) people I need to stay in casual contact with, and (4) people I should not continue a dialogue. The latter would include some of the more dedicated people I encounter at Inventors Associations meetings, who are a little too passionate about their projects and are not generally open to suggestions, sharing ideas or learning.

For those who need assistance with the first category, I am sure you will be able to implement an effective program to encourage these to become clients, in Chapter 12 of this book. The latter two categories need very little attention, but the second category is the key to being able to scale a business. My follow-up letter (included here as a template for you) may appear quite brutal at first glance, but I have found that most readers appreciate the structure and like the idea of "putting in the effort in order to be assured of an outcome".

By using this email, I am able to be sure that if the person I am to meet has read my checklist (and t prepared and sent one of their own). By using this method, I can justify the 10-20 minutes to a coffee shop, knowing that the person I am meeting with is known to me, they have taken the time to tell me what they are good at and the type of referral with which they would be happiest. Before I have set off for this appointment, I would have taken the time to sort through my contacts and pair a client's need with the person I am heading off to meet. My Referral Checklist includes the following details, but would never go over 1-page. It should not have the appearance of a resume or a tender response – just a light summary of who I am, what I want and how to get in contact. Here are the headings I use:

- Full Name and Company
- Location
- Skills and Experience
- Significant Achievements, Directorships, etc
- What Type of Engagements I Seek
- Some Current Clients/achievements I got for them
- Websites
- Best Contact Details.

Similarly, I would expect this person to have done the same type of homework for me and if they do not have someone who needs what I do, then perhaps they are not a person who should be on my networking list. As a word of caution, I do not expect that the referred prospect has been primed or that I will even do business with them, but at the very least I expect that they will have a need that matches my expertise.

So the anticipated outcome for this type of "first meeting" is that each party will bring at least one referral to trade. Given that you may not have an idea of the skill level of the person, it is wise to only bring the one client referral and organise to meet again in say 3-4 weeks with a report for each other, on the outcome of the first referral. This becomes your opportunity to rate the other person's understanding of your capabilities and their ability to match these with their clients' needs.

Scaling Your Networking

There are many business networking breakfast meetings happening all over the World, every weekday morning. Some are highly organised and some are more social and easy-going. The flaw in this model is that the actual referring is always ad-hock and you have to listen to fifty people tell their story in one-minute bites, in the hope that you might be able to think of a client or associate you could refer to some of them. The key to these formats is that they try to broadcast to fifty people, making your message too diluted in the noise of 49 other business people looking for referrals.

Imagine if there were one hundred people attending a business breakfast, once a week, in the same hall. Imagine there may be a welcome speech by the MC and then you were each seated at tables of just 4 people each. Imagine you were given the name of ONE MEMBER, whom you had to sit opposite and that person was allocated YOUR name. Imagine you were given their 1-page networking checklist the week before and you had done your homework by scouring your client list and looking for just ONE referral that you were confident they could benefit from.

Imagine if they had done the same for you and before you presented your referral to this person, you took a few minutes to describe that person's business to the other 2 people at the table (who had each other as referrals and would later present to you each other's business). Imagine if this process, including breakfast, gave you a sound knowledge of the business you researched and offered a referral to, a five-minute summary of two other businesses and you got to listen to a third party present how they see your business, all within an hour.

Would you pay $50 for a breakfast like this, if you knew you were to walk away with a referral every week? What would you care if the organisers paid $20 for the breakfast and made some income for the organising of this event every week…..?

I have yet to see this concept tried, but I do believe it will work. Two of the biggest flaws in networking breakfast clubs are that (a) they are stuck with their members for a year – after accepting their annual fees, and (b) they rarely have competing business leaders in the same club. In the above situation, your membership would be week by week and you only collect a name for the following week, if you are assured of turning up. If you provide dodgy referrals or just do not have the client list or friends, then maybe you do not have the right to sit at a business networking table every week with people who can.

Parallel and Gray Marketing

Two of the more touchy issues relating to market planning are the use of parallel marketing and grey marketing, whether intentionally or unintentionally. For parallel marketing, it is common for home brand products to sell alongside their branded competitors who produce them in large volumes for their supermarket chain customers. Although these generally come in plainer packaging than the standard offers made by the producers or manufacturers, the contents are largely the same. In most cases they have the same specifications and use the same production lines as those of the branded product lines. They can sell for up to twenty percent less as there is no marketing associated with them, and they are sold directly to the retailer in high volume, long-term orders.

The opportunity through parallel marketing for innovators lies in your ability to re-brand your product and sell it into a select, non-competing channel. The market segment becomes more crowded, but you do not miss out on additional sales. You will probably make less money from the rebranded product, but it's best to own this market for the sake of economies of scale on your current production runs.

The grey marketing issue can be a headache, but if addressed appropriately can work to your advantage. Grey marketing generally works like this. A Third-World clothing manufacturer receives your order for one million branded items. That manufacturer produces 1.5 million items and delivers 1 million items to you, his client, selling the balance with your labels or logos. This can be a breach of copyright but it is hard to police, if the owner of the branded item cannot trace the origin of the goods to the factory with which they originally placed the order.

To make grey marketing work for you, there is a way to negotiate with the manufacturer, which is usually in a Third-World country, to offset the right to manufacture additional items for their own market, on the condition that they do not export these to any other markets. This is the generally accepted benefit in the negotiations for production and could deliver a discount on the production price. The key is for them to understand that once they export any of those products into your markets, they will be breaching your copyright and you will seek compensation and sever all agreements.

By giving them unfettered access to their own market, you achieve two key points. First, you give them the opportunity to develop economies of scale with your volume added to the volumes that they will produce for their own markets. Secondly, you do not have to police their market, and given that it is probably in a Third-World country, it may not have stringent rules for patented or manufacturing rights. When you provide your appointed manufacturer with the marketing rights for that country, you are essentially giving them the right to police the use of that product or service or brand in that jurisdiction. It would be up to them to then prevent others from producing copies, which would serve your markets overseas as well.

Giant Client Syndrome

When I started my consulting business, it was the result of an economic decision by a prominent University to close down the commercial arm of their main campus. This commercial entity was responsible for the commercialisation of intellectual property emerging from all the departments across the University. Internally, the vice chancellors made a political decision to axe the department and I went from employee to independent contractor, in less than a month.

I could not believe my luck. I had been given a consulting contract with twelve projects, each funded for the following twelve months by the University. I could work my own hours and didn't need to attend any office if I so chose. The University even provided me with office space, though I chose not to take it because of the scarcity of parking. For the first time in my life I was the boss, and I enjoyed the feeling. I did my fair share of long lunches, but mostly I focused on the work and got the job done. Life was good.

For my first year as a contractor, I had only one client — the University. Within a few months, however, I recognised the danger of this, having been a past victim of political machinations within that institution.

In my free time I set out to diversify my client list, and within two years I had built a sensible and balanced business that was not too dependent on any one customer. I had a marketing degree and two decades of experience in establishing and cultivating business relationships, so I knew this was urgent and important. I eventually spent three days a week doing my actual consulting and one day a week on marketing. This ensured I had a diverse practice by the end of the first year, when my contract was due for renewal.

I suspected my contract would not be renewed, as I saw the investment in my services by the University as a political flag-waving exercise. I felt they wanted to show the tenured professors that expert IP marketing services were being provided for their projects.

At the end of the first year of my contract process, the funding for each of the projects was discontinued. That business I started all those years ago still functions today because I did not fall victim to this giant client syndrome. At the beginning of the first year I had one client who represented 100% of my business. The negotiated terms by which I conducted my practice was on a two-page contract with the Vice Chancellor of Research for that institution. I thought I was free, but in fact I was bound up tighter as a contractor than I was as an employee.

Some of the negative attributes of the giant client syndrome include the negotiation for payment terms and the pressure to be flexible in implementing contracts by the smaller party. As recently as December 2013, I was advised by a client who was providing a substantial range of services to a large resource company, that they had closed their accounts payable for the year as at December 10. This meant any outstanding invoices, mostly due and payable around the 25th of the month, were to be paid in the following year. You can imagine the turbulence this caused for the payroll and other essential expenses which still needed to be paid within their due dates.

This client was organised enough to have a credit facility in place within days, so all personnel received their wages and holiday pay for the summer holiday season. The contractor had to take this treatment, because more than seventy percent of his business came from the single client. By December 2014 this was not the case for this client and they were able to be more forthright with the giant client, to ensure the contract was implemented as per the agreement. Fairness and equitability can disappear when contractual control rests with one party, in this case the giant client.

Another negative attribute is the historical relationship that the principal of the small contract company generally has with the giant client. This may be because in the past, this principal was an employee of that company and had secured the contract as a result of that relationship. Although that relationship is the core reason the contract company exists, it is also the single greatest danger, when the principal tries to do what any good business person does and seeks to offload the core business activities to his employees. In some cases, the giant client simply rings the principal and reminds him that he is the contract, and they do not want other personnel — however qualified or competent they are — covering his duties.

More than a decade ago I was pitching for a strategic planning project with the principals of a major accounting firm. They wanted their strategic planning done independently, and as I had worked with their clients over several years, they were happy to for me to facilitate the strategic planning and write up the business plan. When I turned up to the first scoping session with a colleague, the senior partner took me aside and candidly expressed his requirement that I do the entire project and not have others in my team involved. In his words, "*If we are to pay top dollar, we want the top dog.*" This situation gave me no choice. I either had to take the project on as a personal contract, using my consulting time, or turn it down.

The client was an important strategic partner of my business at the time, having referred many clients then, and recently. I did the project myself and in the process learned a lot about the client, the individual personalities of the principals, and the target industry they had identified as their niche. I also realised I had little leverage with this strategic partner.

Several years ago I was working with a client who was a large, national, listed company involved in providing products to the fast food industry. They had developed three significant advances for the industry, and were looking to establish themselves with one of its leading players. Very quickly they identified that most of the leading players were global organisations, with no strategic decision-making being done in Australia.

We then looked at the list of Australian-owned fast food outlets which were national, had a substantial track record of sales, and could move sufficient product for us. We made our list of the top three players as our preferences, and approached the first one. They were excited about our product. The managing director entered into an agreement with this client firm to independently validate the products, so their competitors could not claim they performed less than what we published. This meant finding independent laboratories and universities to test the products and validate the findings.

This independent validation of the proof of concept took about twelve months to complete, and was partially paid for by this fast food client. They were as excited as my client was to deliver these significant advantages to the general public. Then…, it happened.

Their fast food company giant client, which was privately owned at the time, was sold to a major private equity firm. This firm systematically changed the management of the company, and in doing so, replaced the advocates of our joint venture project.

As a result, our client (through no fault of his own) became a victim of politics and business change, in companies far removed from the original business relationship. Worse still, there was a period of exclusivity associated with this agreement, which was tied to the investment contribution that the fast food company provided. This essentially prevented my client from marketing his range of products to the fast food industry in Australia for several years.

This is one example of **giant client syndrome**. When your business relies on the decision-making of one client for more than seventy percent of your revenue, you are essentially controlled by that client. In this case, a change of management meant that projected cash flows of millions of dollars per year, for several years, were never going to be realised. It took more than six months, and some international clients, to pull this project back into a good place, which has since been reflected in the value of their listed shares.

Transitioning your project to a company is all about scaling the production, to get you more and wider markets. Let us presume that you have run your limited production in your product phase and you know the technology works and has proven reliable. You are confident that your intellectual property is protected, through patents and other methodology and you are ready to take your production from a few units a week to a few million units a week. It may be calculated that your margins will increase and the overall profit now only relies on you being able to manufacture these products flawlessly. You need to scale up to feed the hungry herd of customers that you know are out there waiting for your revolution. How do you do this?

In simple terms, all that you will need is a big pot of money to expand the production, a marketing plan that will assure you of gaining a substantial share of the geographic markets and the range of products or services that can be produced at a competitive price. You can choose to wait until your marginally increased production provides you with sufficient capital to grow the business slowly, or you can turn to a third party to finance the acceleration from this point, to grow the business twenty-fold. This is where venture capital comes in.

Before you are ready to talk to third-party funders, you need to be sure of what you can achieve, if you accept their money. You will now be dealing with very sophisticated professional investment houses that have processes and procedures to qualify your business opportunity, and they probably know your markets and industry better than you do. Once you know you've got your numbers for scaling your business up and you've commissioned and received independent third-party research (which validates your market claims) you will need to focus on what you need to scale the business up.

Your past business plan is almost certainly now redundant. It got you to where you are now, but the next part of the journey is completely different. If you were to maintain your current structure and grow the business incrementally, it would probably be a very nice business to give you a living until retirement, but every year that goes by, your intellectual property value diminishes. It is important, to get the project to full-scale operation as quickly as possible, once you have the production process worked out and the markets validated. You need help to do this, unless you've won several million dollars in a lottery, in recent weeks.

Scaling up

Scaling up a business is a lot like building a race car. If you take a stock standard sedan and you put it on a race track, you will still have some thrills, but you will not taste victory. If you get the world's best sedan car driver behind the wheel, you are going to perform better, but you still might not win. If you double the amount of fuel that is forced into the engine, you will increase the likelihood of winning.

If you were to add a turbocharger to the vehicle, you will also increase the horsepower even more. If you are able to put a racing gearbox in, you will be able to get a higher ratio of gears and thereby increase the speed of the vehicle at the same revs, so your vehicle becomes that much more competitive. Breaking and cornering are just as important, so if you fit a racing suspension so the car does not roll when it goes into the corners at high-speed, your lap times will get even better. Once you put high-performance braking all round, lap times drop even more because you are able to brake later as you go into corners. All of these minor adjustments give you a more competitive car and ultimately a chance of tasting victory.

With economies of scale, we have to tweak and twist everything we do with the production, in order to reduce the unit price of each next product, on the basis that every dollar saved is a bottom-line dollar for the company. Showing an early profit or showing a pathway to substantial profits, is the best way to attract venture capital or private equity capital, to take your project to the next level. In the race car scenario, it costs money to do each of those next tweaks and twists. As you do them, each tweak and twist will drop your lap times and making you more competitive, thereby getting closer to sponsorship money.

Imagine if your project was able to attract a venture capitalist that could generate twenty times the sales and production that you could in your regular growth cycle. Wouldn't it make sense to give away up to half of your company, in order to achieve this? In reality, the half of the company that you give to the venture capitalist as part of their investment, generally does not exist without them.

The advantages that venture capitalists offer you are that they do not want to stay in your project too long. The disadvantage is they want to control their investment very carefully and accelerate your company quite rapidly, before they then bounce out as a cash sale to a third party or a listing.

They start with the end in mind and have no time for emotive attachments to any projects. Although most people see venture capitalists as hardened financial and political animals, the truth is they just do what we do at ten times the pace and have no time for indulgences such as emotional attachment to products or services, or even customers.

Where One Plus One Equals Twenty

Venture capital firms are not interested in owning companies. They recognise that they have money which needs a return on investment. The smartest way for them to leverage that money, is to use their skills and the contacts to strategically build an asset around their money. They generally intend to dispose of that asset within a given period to at least quadruple their money. This can be a good situation for you, if you are looking to partner up with a venture capital firm or a private equity firm, to take your business to the next level

Before you contemplate approaching this level of investor, you need to be clear about what venture capitalists need and what they offer. Most innovators believe that venture capital firms want to lend them money for a fixed amount of time, at a pleasant interest rate, so they can secure a good return on investment for their shareholders.

Venture capitalists generally do not lend money. They simply acquire 50% or more of businesses that they can leverage their money to scale twenty-fold, in order to realise that investment by exiting the company, at a later date. The constant in all venture capitalists' asset registers is the money not the company. They have companies come and go and they simply rotate their money through those companies in order to maximise the return available to their shareholders. You can benefit extremely well from venture capital money, but you can also get burned horribly if you are not ready for it.

Venture capitalists and merchant banks have a different set of investment rules. When you reach this stage of maturity in your company growth, you may need to throw out everything you learned at Cornerstone Investor stage. You are past that.

Several years ago, I had dealings with a project which had an amazing product which we believe could be in high demand by consumers in the recreation industry. The production of this product required some level of technology and learning in a special plastics composition. The company knew they could not produce it in Australia, so they went overseas to an Asian country where they felt comfortable that they could control the production. After twelve months of production and many thousands of sales, they were still only able to produce two products a day and were never going to meet the demand they had.

Their only solution was to scale the project up, using a different factory. They assessed that their primary geographic market was in the USA, so they believed the best way to get a scale operation happening, would be to relocate to the USA. This required a substantial amount of capital to set up a production line, as well as to acquire the components and materials to produce the backlog of products they had.

Our assessment suggested they needed venture capital or perhaps they should approach a private equity firm, preferably one experienced in the recreation industry that they were targeting. It took a matter of hours to identify the top five companies that have invested in this area, and we then looked at the portfolio of investments, to find what we would consider to be the 'dog'. (This term was first coined in the Boston Consulting Group Matrix, in 1968)

This was our interpretation of a large project, with huge promise, that was underperforming for a prolonged period. We were well aware that any venture capital or private equity firm will dump a dog if they do not believe they can turn it around within a sufficient amount of time. We wrote a strategy for this small recreation company, which suggested that they approach the private equity firms that owned that dog, and show how they could turn the dog around, given this new product and the market they had already established with both common products.

Because of the amount of capital required to scale this product up, the current shareholders of the project were reluctant to sell the company outright. This meant the private equity firm could pick up the company in its entirety, back the technology and the intellectual property into the dog, and turn the whole lot around within a matter of months. I was not engaged to negotiate this deal, so I do not experience they went with it.

The key message in that story is to look carefully for the actual investor, based on their pain points, rather than just sending out 5,000 copies of your memoranda to different venture capital firms, merchant banks, private equity firms or angel investors. If you cannot open the conversation in a letter, with your knowledge of their current investments and the pain they are having in those current investments, then you do not deserve a presentation to them and you just become one of the thousands of opportunities that pass their desk every month. In summary, you need to do your homework and you need to show the venture capital firm that your proposal will benefit them well beyond a return on investment.

This may require that you pull apart their annual reports, search for those dogs that need help and prepare a killer proposal that will show the venture capitalists how a strategic acquisition of the technology you have would help the principals to revive the dog. Prepare your pitch around how your project could save them from their current under-performing investment.

What Makes Venture Capitalists Different?

Venture capitalists are not like your Cornerstone Investor and do not share anything in common with your family and friends who may have invested in your project in early days. In getting your project over the second roadblock (cornerstone investment) you have had to learn a set of rules to get your project funded, which proved a different experience compared to your plight to get family and friends to drop money in your idea at the very beginning of this project. Now, as we move to scale up your business we must look to venture capital or private equity funding, which changes the game again. Most of what you learnt in cornerstone investment would be redundant in the pursuit of venture capital partnerships. So what makes venture capital investors different from Cornerstone Investors?

They buy cash-flow; Venture capital firms do not invest in patents or ideas and they most certainly do not invest in blue sky. They are interested in cash flow and if you do not have cash flow available at the beginning of your conversation, there will be no conversation. Venture capital firms play a specific role in the market from the entrepreneur's perspective. They want to see projects which are turning over one to ten million dollars, which can be scaled up to generate $100 to $500 million, given several different factors, such as access to finance, markets and specialist skills.

If your project is turning over less than ten million dollars in revenue, venture capital firms may not be interested in your proposal. This might be contrary to what you see on TV shows, but it is the reality.

They want synergy with other investments they have; A venture capital firm may specialise in certain skill sets, certain markets, certain industries or certain regions. For instance, you may have one or two venture capital firms which specialise in the automotive industry and would locate themselves close to auto markets. They would hire CFOs from these automakers and engage them as analysts for prospective investments they examine. They are looking for synergy for component manufacturers for automakers for instance. If they were able to get a tyre manufacturer and pair this with an automotive suspension manufacturer, their marketing costs will drop significantly and this represents a saving they can internalise, that would not be available to the entrepreneurs or innovators in either project.

They want to match their internal skills with the project; In staying with the automaker VC firm, they would have staff with extensive experience in the automotive manufacturing industry and would therefore be able to lend their knowledge, their experience and their market contacts, to take risk out of any project that they ultimately invest in.

They want to know they have links to established markets; Generally, common products can bring economies of scale to existing markets. If an automaker VC firm was looking at an automotive innovation or project, there is a fairly good chance that they know the market potential better than the innovation team. It would be fair to expect that they should be able to land a long-term contract much quicker than the project team could on their own and their contacts within the automakers would bypass many of the approvals, which could deliver cash at a much quicker rate.

They want to know they can raise the bar twenty-fold; They cannot make money if they just double their earnings. VC firms and private equity firms are not interested in doubling their money. If they can leverage all their know-how, knowledge, finance and contacts, they expect that any business they touch should be able to generate a twenty fold increase from the expectation laid out by the innovation team. They would also experience difficult it is going to be for the innovation team to break through without the contacts they have. This makes them particularly difficult to deal with, as they have the power in the negotiations from the outset. When going in to sit with venture capital firms in such a tight industry, you should expect them to know this and they will use this, every time.

They want control: There is a good probability that the administration team of the venture capital firm has far more experience in driving high growth projects, than you or your team. They know this, and charge you for that service. Unless you have an exceptionally experienced team, there is very little point in arguing your skills are better than their skills in suitability to drive the project through a high-growth phase.

I have included a Venture Capital preparation checklist at the back of this book. You can download this if you log into the website www.commercializeIP.com and click on the link on the front page. Your password is on the last page of this book.

Given that you now have an understanding of what selection criteria venture capitalists firms use for investment, you now need to build your case to present to them once you have an appointment. The first thing you must understand is that these people see thousands of presentations a year and they may become bored very easily. They will become very intolerant of verbose or passionate presenters who want to branch out on their own tangents or run down their competition. They are interested in a very tight set of criteria and most importantly they are looking at the presenters to see how well they can perform as a cohesive team. They want to see how cohesive the project team is in a high pressure situation, such as when presenting for venture capital.

They will almost certainly have Googled the individuals and the project before you walk in and may still be on your LinkedIn page while they talk to you. They may question you about contacts in past companies. If you are from an industry they are familiar with, there's a very good chance they know key people in your past firms and want to make sure that you have positive connections with those people.

One of the first sins that project teams commit is death by PowerPoint. It is important not to labour every word on to PowerPoint presentations but rather provide details in a handout at the end of the presentation. They want to hear about the significant key differentiators you have and all the compelling reasons why they should be part of your project forward. If you are not able to do that in the first five minutes then you should enjoy the cup of coffee they offer because that is all you are going to get from them.

Venture capital firms will have a structured set of questions after your presentation, to learn more about you and your experiences. They will want to know about all the team members, why they are included and more importantly what projects you've handled in the past. If you talk about how successful you have been with a certain widget, they will ask you to expand on how you made that successful, while they are Googling the project, the company that now owns it, as well is the value of the company before and after acquisition.

If you had past failures, there is no point in hiding this. Venture capitalists spend a lot of money with analysts whose job is to determine the risks associated with doing business with you. It is fair to suggest that a lot of venture capital firms have just as many failures as you and one of the biggest failures you can have is hiding a bad history.

The most critical point to remember in dealing with any sophisticated funding providers, is that you have to know exactly how much you need, and what that will return, before you ask for it. If for any reason, you run out of money before reaching your cash flow positive situation, you will face very heavy consequences in losses of your equity. If a venture capital firm were to place ten dollars with you for a project, and you need another five dollars to make it work, consider that the second five dollars will cost you at least four times what the original ten dollars will have cost.

Preparing to Scale Up

By now, you have commissioned your market research and you know the demand for your product or service. You have completed your prototypes and run small-scale production runs, to prove you have consistency in production and reliability in the field. You are now ready to take your production run from say, 100 a day, to twenty thousand a day. Scaling up can give project team members the biggest thrill of their lives, but the task itself is quite daunting.

Scaling up for a significantly larger production run brings a completely different set of issues, most of which you and your team may not have experienced before. For instance, you may have calculated the space you need for a production plant, but when the production gets into full swing, you need to have space on the floor for feeding components in at different stages of the production facility.

This is not a problem when you are producing 100 a day, but when it scales to twenty thousand a day, you really have to know where components are going to be placed and used, so that just the delivery alone does not tie up your workforce.

The next consideration will be resources, such as power, water and other services that you will need to operate. These sorts of calculations involve specialist quantity surveyor and/or production systems engineering issues, so we recommend that you factor in a consulting engagement by a specialist engineer, who can scale your project in increments and still deliver you a healthy production output as you grow.

The next part of your preparation for scale is going to be finance. I recall a small production firm in Australia, who developed a washing machine for engine parts, to be used in the automotive industry. This little four-man shop scored a contract with a major auto retailer with some 180 branches across Australia. In order to supply these, the guy scaled up his production for these machines, which cost him $4000 per unit in production and components. The auto retailer chain wanted to lease these machines from him at a very attractive rate, over a four-year period for each machine, at each location.

However, this guy had no idea of how to scale. He could produce up to 12 machines per day, but with production cost of $4,000, he could not offset this with the income from his rental agreements. His solution was to bring in equity partners, which would dilute his share of the project, almost down to nothing.

We structured an independent funding facility, which purchased the lease agreements from him, after he signed to place the machines into each of the outlets. This funding facility collected the lease agreements over the 3 to 4 years and cashed him out of the agreements upfront, at an agreed discount, within 48 hours a placement of the machines under contract. He was able to receive almost $9,000 in contract prepayments, for his $4,000 machines and was able to own the machines after the lease expired.

His downside included having to either pay the lease agreement if the contract fell over, or place these in another location so that the independent funding facility was able to receive the same revenue. Given that he usually deals with major international car dealerships, his failure rate on contracts was going to be very low.

In many cases, the innovation teams do not enjoy scaling up a project, once they've achieved the funding. It is fair to say that this is a completely different skill set required and in most cases, the innovation teams would like to get back to innovation and leave the production to professional production managers.

Rather than expect these people to leave you and simply replace them with a new team, it could be more prudent for you to plan for these people to return to your incubation centre and offer them the next shiny box to be opened. If you plan to have a different team for the scale up, you will have less difficulty in transition by people with experience and knowledge mostly lies in development, but not necessarily production.

CHAPTER 10

The Value Realisation Point

"Start with the end in mind"

—Stephen R Covey

You've reached a point where you have put your project into commercialisation, built a company around the production and distribution of the project and now are looking to realise that payday that you promised yourself. In earlier chapters we asked "What is the best time to plan your exit?" To save you turning back to previous chapters, I should remind you that the best time is at the start. However, if you have got this far without an exit in mind, it is never too late to start the planning.

Firstly, you need to know what your options are to exit the project. This may or may not mean you getting out of the project or getting out of your company. You then need to look at what you are okay to do beyond this all-consuming IP commercialisation pathway. In chapter 2 I introduced you to the IP commercialisation pathway as a circle.

The final point, being the exit or value realisation point, allows for transitioning from the current project to a new role in another project. You may have many of those intellectual property ideas that we asked you to park in the first chapter so that you did not become confused with ownership budget allocations.

Now you can unpack those shiny boxes, and determine whether any of those is equal to or better than what you have and worth the process again. In many cases you may have picked up the most attractive one at the beginning and now that you know the process better, you may decide that some of the others and not attractive and may not even be worth commercialising. In that case you may be searching for a role for yourself beyond a value realisation point.

You may recognise that the trade sale of your technology will not give you an ongoing role in the project and this may create some reluctance in the value realisation process. To keep yourself involved you will need to have unique expertise to bring to any other R&D project for that matter in the commercialisation space, now that you have completed the first of what could be many journeys. The most obvious role for you going forward in new projects, would be as a Cornerstone Investor.

You expect that at the end of this value realisation process, you will have some money available for investment. You may also have developed some unique and positive expertise in the commercialisation process, which you can share with innovators in that space.

To recap the role of the Cornerstone Investor, it is generally best suited to somebody with a passion for that business, the industry and/or the model. Most Cornerstone Investors may also have a company which supplies that industry or market, in which the new technology could leverage a greater market share or enable them to command a higher price.

So a Cornerstone Investor brings cash, expertise, market-knowledge and commercialisation know-how. They can mitigate their risk because they understand their industry and respective markets. This investor can understand where novice innovators will waste resources and they are highly attuned to the cash burn in any early-stage project development.

They bring much more than money and may play a pivotal role from the outside. In most cases, they are given a seat on the board and will generally provide their investment on a monthly drawdown basis, against agreed milestones. For most innovators who have completed the process, this becomes an ideal role, where they can make a significant difference in many projects, for as little as $500,000 each investment.

The value realisation point is a time when a transaction is completed, which vends the intellectual property and the associated project, to a third party. This third party may be a public entity, a trade sale buyer or even the commercial arm of the same company.

As with the trade sale audit performed at the IP handover, back in the transition from concept to project, there is a formal process involved in assessing and handing over the project in its entirety. This can sometimes take weeks and at other times can take years. Depending on the complexity of the project, the principal needs to be aware that he may be required well beyond the value realisation point, in order to help the trade sale buyer to realise the full value of the purchase. This may be in everybody's interest, as the vendors generally have a performance component to their trade sale price.

Valuation for Trade Sale

In September 2017, as a favour to one of the members in my online INVENTORS MASTERMIND group in FaceBook, I jumped on a conference call with him, to initiate trade-sale or licensing talks with a North-American prospective buyer, who happened to be the manufacturer of a range of products my client distributed for them exclusively in Australia.

John (his real name) needed to justify a sensible price for his IP and was stuck on the only valuation method he understood – what it cost him to build it. We talked about other methods and I shared with him the 5P model for maximising IP value for licensing or trade-sale agreements.

Some years ago, John had identified a need for a product that would optimise the current range of products he was importing and distributing across Australia. He was convinced this new technology would help make this existing range of products more competitive, in technology, reliability and price. His "black box" device eliminated the need for this very expensive 16-core copper cable in between the units (one site has 11 kilometres of this stuff – at $8.00 per metre!) and understandably, the users fully understood they could save on price, shipping costs, installation and maintenance - if all that cable wasn't required.

We had to build a case for why the manufacturer needed to acquire this IP for their range. We did our research well before we got on the phone to speak with them.

Firstly, we identified that they were wholly-owned by a publically-listed Company, with approximately 40% of that stock being owned by a private equity firm. This was all good news for John, because we could ultimately negotiate a cashless transaction, by simply swapping the equity in his R&D company for shares in the public company.

This means the transaction doesn't affect cashflow or the profit and loss statement of the buyer and where the inventor operates in a tax regime that features capital gains taxes, he may be able to forego these gains until the moment he liquidates the new shares he has acquired.

Private equity firms understand the value of a cashless transaction and don't mind the slight dilution for adding a cash-generating asset to the portfolio. It's an easy win for them. To structure this in the most attractive way, we applied the 5P pricing formula for licensing/ trade sale negotiations. Here's how it works

Pricing

The first stage of winning the best price possible is to build a value model from the perspective of the prospective buyer. Some of the questions you need to ask yourself include:

- What could the buyers sell these units for?
- What could they produce them for (given they will have better economies of scale that you)?
- Can they access the resources and facilities to produce these?
- Is the competitive advantage less for national and even global exclusivity, than for local exclusivity? (it might be more efficient to segment the licensing or trade sale to a territory and negotiate multiple licensing agreements)
- Could this make their existing offer more competitive?
- How many of these units could this buyer go back and immediately offer to their existing or past customers? (easy sales, plus reengagement)
- Could this be used to get them new customers? (Can we calculate the value of each new customer?)
- Could they use this innovation to take customers from their competitors? (by promoting their "complete package" as a new market positioning)
- Could they take distributors from their competitors' networks? Surely most of the competitors distributors will hate to lose sales because they do not have access to something that ultimately might be featured in the eventual tender. (This would surely shift or enhance the existing Industry dominance).

This assessment becomes the first qualifier for who should buy the technology or licence. If you can present the business case (for acquiring this technology) from their perspective, you might later overcome one of the greatest stumbling blocks in the process – getting to the right decision-maker. Sometimes, you cannot get in front of the right decision-maker and if your contact presents the idea armed with nothing but enthusiasm, it might get shot down before you get started.

Perspective

The second stage to valuing the asset is to measure their opportunity cost of DIY. How much labour would it take for the prospective buyer to do this themselves and how many months/years could it be achieved in? Is this less than what the current product could be established in the market for? They need to be reminded that if they didn't buy this, who would and what would that do to their cashflow, if their competitor had the advantage? The answers to these questions can be modelled and measured by a good accountant.

Modelling this option becomes critical if you are not present during the discussions. You cannot expect to have everyone in their evaluation team to be positive about this acquisition, so building the rebuttals to un-declared arguments could help those more enthusiastic advocates to justify their position with cold, financial logic.

Paperwork

The third phase is compiling the due diligence file. This means sourcing as much external information and secondary reports from unrelated sources, that support the conclusions you made in phase 1. For John, this meant taking a large, international tender which he bid for (and didn't win) with the manufacturer's equipment, and show how much more competitive they could have been with this device.

It was easy to identify a saving on this one project, of nearly $250,000 in removing redundant equipment, the reduction in installation costs and the price of 5-years maintenance (which was part of the tender). Having an actual tender document restructured using this technology, immediately showed how they could overcome the first issue in most tenders – price.

Without stating the obvious, this will also show the prospect just how much cheaper and more efficient their competitors could be, if these decision-makers pass on this project and the competition takes up the option.

The Players

The final phase is to get a better understanding of with whom you are going to be dealing. We profiled the key personnel within the target company and we researched the ownership up 2 levels. We understood this could go well beyond the pay-grade of the local management and we didn't want to worry if they had the skills to present a winnable business case. They made it very clear they wanted this, but we know it was going to be up to us to help them get this done.

Before we offered this product to them, we asked them if we could help their distributors become more competitive by letting them know we had these new units available. They agreed and provided us with a list of their global distributors. We were able to write to these distributors and introduce ourselves as part of their global network,

As part of this marketing program, we gave each distributor an opportunity to receive a pair of these units to evaluate. They had the option to return these within a few months or simply pay the accompanying invoice and on-sell them into their next project. There was little risk because of the longstanding relationship they each had with the manufacturer.

By the time we got to the negotiating table, we had a persuasive argument for why the prospect needed this to remain competitive in a technology-evolving environment. We knew the people who would be seated across from us and we knew the people they would have to present this business case to.

We had prepared an encompassing presentation on the technology and the business case for acquiring it, as well as a detailed expansion of the "do nothing" option. As this industry was considered a duopoly, we knew we could play to their fears of their larger competitor being able to access this and use it in the same way we proposed they used it on them.

We made it understood that they had the first option, that if they passed they knew where it would be presented next and finally (and very importantly) they understood that the decision for taking it to competitors was not John's. This took all pressure off John for the duration of the negotiations (the technique is known as "vague higher authority").

Payment Options

Pricing the intellectual property was easy for us, because we had experience in this area. For the novice, this is the hardest option and can result in selling out too cheaply or not getting a deal because of affordability. On the affordability issue, we asked ourselves one question:

What if the buyer could acquire the technology for no cash at all?

Our research had identified two important facts. The first was that 100% of the shares in this manufacturer were owned by a publically-listed Company. The second fact was that this publically-listed company was 40% owned by a private equity firm. This was good news from the level of sophistication we could apply to this deal, but bad news from the perspective of how experienced and powerful the negotiations would be.

This simply meant John would need someone experienced in his corner and he already knew he had someone talented and experienced to turn to – yours truly!

We structured a deal which would enable the listed parent company to issue equity in the main company to the value of the price we had decided was fair and justifiable. These shares would be tradeable and would be ordinary shares with all entitlements (no special class of shares).

The transaction was absolute in its structure, but gave the vendor a buy-back option for a fixed price, if for any reason the shares in the Company dropped more than 30% or were suspended. The buy-back included a "first-right of refusal" (at equally-favourable terms) for any trade-sale of that technology to a third party. We tried also to negotiate in a 5% share of any licensing agreements the buyer did, within the first 5 years of ownership, but this became a concession that was negotiated out in the process.

To make the deal even more attractive, we had a sceptics' option. This allows the buyer to be absolutely certain the project is a good one, by reducing the initial issue of shares by 25% and offering a further 60% of options in the Company, exercisable at the current 90-day average price. This ensured the vendor could not benefit from the entire sale price, unless the Company share value increased. This would be inevitable, given the value of the technology.

There are risks associated with a sceptics' option. The Company circumstances might change for reasons other than the project and this will affect the options value. The Company might sell or trade the technology to a third party and offset the transaction price for debt or some non-cash asset, which might not complement the share value.

The underlying assumption one has to make is that Directors will work to increasing the share value for all shareholders, as part of their fiduciary obligations. Most of the time you can review the published audited accounts with the relevant stock exchange to know how risky this option can become.

Because the ink isn't totally dry on any agreement at the time of printing this book, we cannot publish financial data on the deal. We can confirm this was a mutually-satisfying transaction and that all parties are very optimistic about the future of the technology involved. John hasn't booked his holiday yet, but he is looking at monster caravans to make holidays a far more important part of his working life.

Profile of IP buyers

If you are intending to conduct a trade sale as your value realisation point, you need to select the preferred option, which is not just around the size or timing of the return. It could be that one of the more important factors to you is the ongoing development or the role you may have in the trade sale buyer, or even retaining access to the product or service, for your own market. Your more common options would be to trade sale to a large organisation, to list the product or project onto a public stock exchange, to vend the project into an existing public company as a reverse takeover, to license the technology to one or many players in the industry or to enter into a joint-venture program that will deliver vertical integration of services or products.

The least desirable option in most cases, is to grow the project out. Unless the parent company has access to a substantial amount of funds, it will be very hard to grow the project out from nothing. As the project gains momentum and increases in sales value, it attracts competitors who may not be in the same geographical or demographic space, but may be attracted to it by the profits. They then become immediate competitors and, in some cases, may reverse engineer your project or process, to beat you to the markets that they control.

One of the more prevalent "grow it out" options is franchising. Most franchises require protection on their element of uniqueness, but also will require proof of any potential, through a flagship store or outlet which has traded for at least three years. In that time, you are announcing to others that this technology is out there and if you do not choose to license it to anyone else, they may try to circumvent your placement or other protection. So starting a franchise from scratch, can be treacherous.

For the same reasons, a multi-site growth program, requires the same or more cash and delivers only incremental reward. The incremental returns would have to be reinvested in new markets, which would prevent the principals from earning a dividend on their projects for many decades to come, if they were truly committed to growing the project outputs for potential.

Having presented calculations of options to intellectual property owners in the past, I have been able to establish in most cases, if they were to grow the project out post vending to a listed company, the exponential returns would outweigh the equity they give away.

Because there is such a commanding justification for other options against growing out, it becomes evident to any outsider that a grow-out program may be focused on the ego of the principal. In chapter 1 I presented the five key requirements for commercialisation by inexperience inventors.

These included a building in the city with their name on it, one million dollars in their bank - not the company's bank - 51% of the shares going forward (after the deal is done), their friends on the

Board of Directors, all the money they put in to be repaid as if these were loans and finally, a global fact-finding tour for themselves and perhaps their families.

As unrealistic as this appears, it is very common starting point for a lot of inventors. By now, you've been through the process and understand the hard work you put into this and as a result the project is worth considerably more than what was as an idea. That does not make it worth all of this and it never will. The only people who will pay that sort of price for your technology would be your parents.

The value realisation point begins with a transaction. This is a process of identifying and the negotiating with the desired parties, to either share or buy the innovation, once it's in high-growth mode. As with commercialisation itself, the trade sale requires a completely different set of skills from R&D.

Don't believe you can do this yourself. If you believe you can find $100 in value within your intellectual property, I could guarantee you someone embedded in that industry could find you $1,000 and is not it better to share even half of something that is worth more than ten times more? The most popular trade sale mandate would feature a process to share everything over the minimum amount that you believe is practical, as an incentive to the agent, to get a higher price.

If you are intending to enter into licensing agreements, then the consultant you engage to do this needs to be aware of this one single fact. The more licenses you sell using geographic and/or demographic segmentation, the more revenue you will yield for your IP value and sales efforts.

If you have been reading this book from the beginning, you have already formulated your exit. However, let's for one moment assume you are perhaps standing in an airport bookshop and flicking through this, when you zero in on this chapter as being of the most relevance to your project at the present time.

At this point, it might be prudent to remember why you built this project, was for one of two core reasons. Perhaps your intent was to realise a substantial passive income from the idea, or perhaps it was to receive a massive windfall from one or more willing buyers. Even though the project should be at a point where it's turning cash and growing at a very rapid rate, it is still not providing you with a sufficient yield to grow the business out at a competitive rate. Any thought to grow a business out, including franchising and/or multi-store growth, could be optimistic in its planning. In more than 90% of cases, the second option (and all of its derivatives) would be the attractive one to an IP project in high-growth mode.

If you have decided to realise the value in your project, and you have selected the best method to do this, you will need a facilitator to make this happen. The most pivotal of advice I can give you for the price of this book, is that you should not put this task in the hands of a real estate agent or business broker.

To speak in general terms here, most business brokers do not understand how a valuation on intellectual property is done. They would stick to how they value small cottage businesses, which is a multiplier of profits or turnover. When you are in the high-growth mode, your turnover from last week will not in any way resemble that of next week and may be less than 1/20th of the equivalent week next year.

Any traditional valuation method is going to undervalue your project substantially. You need to engage an experienced established consultant, to the industry you are targeting.

Find a consultant who is providing services to the senior management of the target company (or companies). You will want to see that they have been engaged by one or more of the companies you are targeting in that industry and have a rapport with senior management of these companies. It could be dangerous if this consultant has an arrangement with just one company, as his negotiating position may be biased towards the buyer. However, having the right incentive program in place for him, should swing that back in your favour. Let's look at the different options for the value realisation point.

For Trade Sale

- Find your target first. To best orchestrate a trade sale to a third party, you need to have developed a careful profile of the company to which you would want to sell. It may not be a specific entity, but will be somebody in that industry. It is always good to pick your target and build your trade sale proposition around their needs.

- If the target entity is a public company, they do not have to pay you cash for any merger or acquisition. It would be more prudent for you to accept equity in their company in exchange for equity in your company, so that (in most jurisdictions) you may not face a sales tax, capital gains or other tax event. If you are simply exchanging shares on a merger, it can be considered business as usual.

- In the next chapter, we talk of running an incubation of many intellectual property projects. In there, we mention the need to have each intellectual property project separated into its own corporate entity. This enables you to exchange equity for equity, without having to include or exclude any other profits or losses associated with other projects. If your intellectual property is a subsidiary of your own operating company, it will still pay to have that owned by a separate corporate entity, to give you the flexibility to sell this without affecting the other company. Nobody likes to have to conduct due diligence on the established company, because they're buying some intellectual property and need to make sure there are no skeletons in the corporate closet.

- To estimate what this company might pay for intellectual property at acquisition, you may be able to calculate previous acquisition values from their published accounting statements (they always say the same method of valuation, that is to satisfy the auditors) that "this IP is worth a factor of the future sales potential, once it is owned by them."

- If you study the business model carefully, you can also factor in additional sales they will get from their existing product and service lines, once they include this intellectual property. There may also be an opportunity for them to retrofit client installations with your intellectual property, which may in itself, generate a windfall.

- You could also calculate the value of their relicensing of the new intellectual property, into other jurisdictions they do not operate in. This service can mitigate their risk and may even repay the purchase price several times over. Although this option sounds lucrative, is not always possible for small companies to achieve this.

- Your summary of value should conclude with a HUGE discount for risk. Naturally, engaging you after the sale would reduce that risk, but that is up to them.

- Effectively, your pitch should be what they need to sell this to their shareholders. This is your chance to deliver a "made-for-you" proposal for their next AGM or circular resolution of directors, to approve the acquisition.

- Consider that not all the decision-makers will be in that room. Most of them might never meet you. Your pitch needs to cover that.

- When you offer a pitch deck/handout, tell them you will send it in a few hours, because you want to include their questions and your responses, as an annexure. This helps answer any questions from those who are not in the room.

For Joint Ventures

- Your pitch to potential joint-venture partners must summarise what you bring to the table and what they bring to the table. This should be done immediately after product description and market analysis.

- You need to address the risks for the joint-venture partners, including their liabilities if this goes wrong. It could be that your project can deliver a product or service to their existing customers and their risk is associated with what those customers would do if your product or service did not deliver.

- There are many intangible benefits for joint-venture partners, including enhancements to existing networks. A case in point is a small electronics distributor I have dealt with for many years who were the country distributor for a global supplier in an industry with just two major manufacturers. Our pitch to the company that they were dealing with was that we could enhance the sell-ability of their products, to their entire dealer network in every country. They gave us contact leads for 44 countries and invited us to approach these distributors and offer that product for them. Their only benefit in the transaction was the suitability of their product to this innovation, which combined, would reduce the sale and installation price for many of the major projects. They helped, because this would help them sell products and there was no risk from their perspective.

- Any joint-venture arrangement has to have an exit. It could be that you plan to run this for two or five years and either list the project or agree at commencement, on a trade sale price, based on the multiplier of sales at the end of that period. Going into a joint-venture without an exit is handing control of your project to somebody without any guarantees. Even if that joint-venture agreement has a minimum sales arrangement, it could be that your product gets shelved and you just received the minimum payment, as a way for them to buy their way out of competition. This becomes a win-lose situation.

For Listing

- Listing refers to preparing an initial public offer of a public company to be listed on the stock exchange in your jurisdiction. There are several different levels of stock exchange in most countries, and for technology, it is always better to look at the lower boards. These have more speculative investors and will take early-stage underperforming projects, before they grow and graduate to the main stock exchange board.

- One of the more creative ways to list a company, that is not available in all countries, is a reverse takeover of an existing public company, where that company has been under-performing and may have lost its core revenue-generating assets. As a result they may have a few million dollars in cash reserves and no assets to generate revenue in the future.

Their only way of survival is to find new technology or new products. These are handy, because they firstly are already listed and that may save you up to half a million dollars in listing fees and specialist reports. They also have a fixed value being the listing fees plus the cash, but the downside is you may have a lot of legacy shareholders and some directors, who do not wish to participate in your project and may actively sabotage it long-term. It is important, in these circumstances, that you get at least a significant share of the company by negotiation, to protect yourself from hostile interest groups.

- If you are contemplating a reverse takeover option, you need to be aware of the sizing. This is where an independent accountant or other expert looks at existing legacy shareholders and your market valuation, and determines how much you should be allocated and what percentage of the company that will be at the end of the relisting process. Undervaluation can give your very low returns, and you should have it in the context of the relisting value, not the current value of the company. In most cases, the company needs to raise a substantial amount of capital, before it re-lists. This required capital will need to be included in the sizing exercise, so you know exactly what it is you have in the share in this relisted company.

- If yours is the only technology being vended into this company, it is generally accepted that you receive some board nominations. You should be aware that if yours is the only product in the company, that you should at least have half the shares and that you should nominate the chairman as an independent non-executive chairman. Some people accept half of the directorships, but end up accepting the existing chairman. In some cases, he may have colluded with the other directors and deliver majority control to them. You should also be wary of existing directors who offer to find an independent chair. If this chairman proves to be partial, you are effectively in the back seat of the bus that you own.

Aggregation

- Aggregation is a selective compilation of intellectual property and other cash generating businesses, who share common markets, may sit across the same industries and who need the same technical service and support.

- Bundling groups of trading companies into a listed entity can be fine, but you should be aware of the need for a balance of cash generating businesses.

- If you have your R&D in a public company you will be generating losses and your share price will be decimated by the market. This will reduce the value of your assets significantly.

- Aggregations are dealt with in greater detail, in a later chapter.

Licensing

- The fastest way to risk-free cash in the realisation of your project, is to negotiate a licence for someone to manufacture and/or distribute the product or service. Traditionally a licence would include a modest upfront establishment fee (commonly called a signing fee) and some percentage of sales going forward. This can be quite lucrative to get large partner companies moving, without them investing substantial cash which may require shareholder approval.

- One of the keys to licensing is accountability. If you have negotiated a licence fee as part of an ongoing portion of sales, there needs to be a way for you to measure this. The best way is to calculate what you believe the sales will be, and fix a quarterly licence fee against the minimum amount of that, so that you do not require auditing or other scrutiny of their sales or business activities.

- With a fixed quarterly licence fee (commonly called a royalty) you might factor in a discount for the first year, in order to help the licensee establish their territory.

- The key upside to licensing is that you can segment the market in many different ways. Traditionally people want a licence for everything and everywhere but the reality is, if presented correctly, they will settle for a geographic or demographic licence.

- This may cover more than 90% of the territory they are operating in on a geographic or demographic basis. This gives you the opportunity to sell the licence to other territories and to other market demographics. You simply place a limitation on each geographic or demographic licence territory.

- The downside to licensing is that it may be vulnerable to exploitation. In most cases, the licensee is substantially larger than the licensor and may abuse that market power. This could be through under-declaring their sales or worse, through a grey market manufacturing program. There are legendary stories of athletic shoemakers, who commissioned a factory in Indonesia to build 100,000 pairs per month and found that the company was ordering components to produce 140,000 pairs of shoes per month.

An audit subsequently showed that many of the extra products were being sold on to the Indonesian market, as genuine branded shoes, without any licence fee being paid. The audit process to prevent this, would be very costly for a small company. We recommend that licensing be done post-listing, where the IP owners have deeper pockets and can defend any breach of patent, licensing or distribution agreement.

The Dash for Cash

Once you have finalised the exit model, you now need to establish how you will build additional value before the sizing or valuations. This becomes a mad dash to the finish, because everything that you generate in revenue from market agreements will lock in a better final valuation, which will directly benefit your sell price.

The first decision you need to make is "are we building value or sales?" The extreme examples to be used here would be Facebook on value, where they generated no revenue but were still worth billions of dollars. Another example would be where a technology generates fees on an ongoing basis, and the valuation is based on the net present value of future earnings, discounted for the risk. This would require two separate tactics to increase its worth at valuation. The first would be to increase the number of members and the subscribers who use the service. The other would be to generate more sales contracts, particularly with future cash flow attached.

Having a comprehensive marketing plan which will show from that day forward that you can generate a substantial amount of cash, will also prove value to the independent accountant or valuation consultants.

At this point, you will already have created a revenue generating machine or marketing department, and in this phase are about to accelerate this with everything you have, because it will reflect on your final price. I do not advocate fudging figures or writing marginal contracts, but doing what you would normally do in two years, compressed into the next twelve months, could quadruple your value when you apply a multiplier. Is not that worth it?

CHAPTER 11

Owning the Process

"What got you to here, will not get you to there…"

—Marshall Goldsmith

If you followed my process so far, it should have helped you navigate that awkward space between innovation and income. Now you are presumably out the other side, you have forgotten the pain and disappointment of your journey and you suddenly miss all of those sleepless nights when it looked as if you could lose your house and as a result of this stress, your family all hated you for a time. The most prominent memory of those emotional-rollercoaster times, are most likely based around the thrill of creation and the pride in knowing that many people around the world use something that you developed.

The most sensible compromise is to enjoy the thrill of building something from an idea into a fortune, but without the rollercoaster ride of financial and emotional hardship that is commercialisation. There is another option.

The easiest way to keep your sanity is to participate in projects conceived by others. In most cases, our graduate innovators tend to come back in and invest in other projects that we have, to keep the thrill going. The key is not to own the project entirely and not to fund it completely, so that it cannot drain your cash and ultimately the life right out of you. Most innovators would prefer to have just the one commercialisation process in their lives and try to continue in an independently funded environment, so they do not burn out.

So if you are ready to climb on the bandwagon to go around again, there is a sensible way of doing this that will not drain your funds and will keep your family talking to you for years to come. This is to become the Cornerstone Investor in someone else's project. In the past 28 years, I have had many innovators come back into the system and become investors in new projects, once they have a firm understanding of how this process works and how we manage the risk for them and for others.

If your recently completed project has yielded you a public offer or a trade sale to a public company, you will not want to compromise on small innovation, but rather you have seen the opportunity to build massive wealth from arbitrage of innovation. More specifically, you might use your commercialisation experience in packaging this innovation specifically for public listed companies. There may be fewer thrills in the process, but the financial rewards are still there.

We call this process "aggregation" and I deal with that in greater detail, later in this chapter.

The Cornerstone Investor

In earlier chapters, we spoke of Cornerstone Investors for commercialisation and project development funding. Typically, a Cornerstone Investor has a passion for the process, understands the industry and the market intimately and can benefit through leverage by owning a piece of the intellectual property project. They rarely do this just for the money, but money is a very important part of sustaining their interest.

I suggest at this stage, if you believe you could become a Cornerstone Investor for another project, you go back to the earlier chapters and revisit the role, the attributes and the benefits of a Cornerstone Investor in the commercialisation process. This position becomes the project champion and lends objectivity to any intellectual property commercialisation process. Rarely does a Cornerstone Investor become blinded by emotional attachment or loyalty to a product. This level of objectivity can save projects from massive disasters, because at the very heart of this role there is a degree of conflict and mastery that every IP commercialisation project needs.

Becoming the IP Magnet

As a potential Cornerstone Investor or commercialisation facilitator, the expectation is that you will have two attributes that every innovator is looking for. The first is that you have survived the process and prospered, therefore developing experience that can be shared with others about to embark on the journey. The second is you have funds or access to funding, as a result of your past commercialisation process. You should never underestimate these attributes and the demand for them with almost every innovator in the world.

But the skill which sets you apart from all others is your ability to incubate projects and make the budgets last for the entire length of the project. You got through a process that many others will never do, regardless of the value and benefits of their projects.

So the best position that you can take in the future, is that of the incubator. To do this you have to set yourself up to become an IP magnet. This means that before you start to attract any of these ideas and their colourful innovators that may be championing them, you have to be able to establish a filtration system to prevent you from becoming bogged down with too many ideas and that can weed out those that may prove too difficult for you to incubate.

The first stage of becoming the IP magnet is to have your primary filtration process established. This means you need to know which type of projects you are looking for and just as importantly, which projects you do not want. If you understand that the majority of seed capitalists and venture capitalists say no to more than 90% of the projects that are pitched to them, then you get an idea of just how brutal you have to be in this early process.

Where Do You Fit?

The first stage of becoming a successful IP magnet is to understand where you sit on the innovation cycle. Most of the innovators you meet will put you on the venture capital stage, hovering between product and company. Your own internal assessment should be based on your skill set and where you most enjoyed being as you transitioned your project through the early-stage commercialisation and into the value realisation point you achieved.

Ultimately, the best resource you can bring to any project is going to be experience and enthusiasm. Money will be required (and make no mistake, almost every innovator believes that's all they need) but projects need more than money to become reality.

Where you see yourself on the innovation cycle, together with your skill sets and your industry and market knowledge on certain topics should very quickly shape your profile as an IP magnet. Once you have defined yourself you then have to establish real filtration systems because the moment you start to promote availability, you will become overwhelmed. Most of the inventors who approach you are not likely to take no for an answer because they have read all the motivational books that say success is only achieved by persistence after everyone else has rejected their pitch. As inspiring as these words may have been for you all those years ago when you start out it could become the most hated phrase you hear in your early future, if you do not have a robust filtration processes in place.

The Filtration Process

In earlier chapters, we alluded to the five key points that all venture capitalists used to qualify early-stage commercialisation opportunities. These were stated as (1) the people, (2) the product, (3) the industry, (4) the markets, and (5) the people. By way of revision, the earlier chapter explained that "people" appears twice because in every project you will look at in the future, the commercial viability of it will depend on the people. The viability of all projects starts and ends with the people. This is not about people you know, people you can trust or people you are related to, but rather the attributes of the people involved in the project and how readily they can put aside their egos or their need to control, for the greater good. By way of revision, I remind you that the product aspect of this refers to the business model, with the asset playing a support role.

The next big filter is geography. It pays to only look at projects close to you, because the downtime that you will experience moving between projects or from working on projects to being with family etc., will wear very thin over time. If you have to spend hours each week or month in aircraft or on highways, you will ultimately turn to thinking this project is starting to sap the life out of your otherwise perfect existence. Even though telecommuting has eliminated a lot of these difficulties, you will need to be face-to-face with all the people in your project on a regular basis, or otherwise there will be misunderstandings and even outright deceit, which will drain the resources out of any project and, in particular, the money out of your pocket.

Next you have a demographic segmentation to decide. When you become the IP magnet, you will attract very nice projects in which you have absolutely no expertise. A case in point is software. In the 1980s and 1990s, the tech bubble burst because the investors jumped into projects not understanding what the technology did. They relied on the R&D people and were not always given all the facts. They understood that some projects, such as Microsoft and Apple had made billions of dollars in short periods of time and they were blinded by the opportunities that software development could produce. It is fair to suggest that more than 99% of those projects funded by venture capitalist at the concept stage, never made it into production and certainly never made a dollar for the investors.

The lesson here for you, as you become that IP magnet, is to only pick projects which align to your area of expertise. It may be that your experience is in the marketing channels or the industry relationships, but that would still be specific to one category of project and if you set your filter to only allow these projects through, you will not have to worry about understanding the technology in front of you when people are asking for more money.

The next filter is size. When you are looking at a project that requires half a million dollars to commercialise, there is a fairly good suggestion that that will ultimately need about $1.5 million before it starts to turn revenue. If your pockets are not that deep, then you should be very careful about starting a project you cannot finish.

You can imagine that as any project begins to run out of money and is forced to raise more capital in the concept or project phase of the cycle, and all existing shareholders, including yourself, will have to give up some equity to attract this new investment. The temptation will be for you to chase your own investment, to ensure you do not dilute your equity ratio in the project. The effect of dilution would be to create the same anguish that you experienced during your commercialisation process. To better manage any project, you should only look at an investment that is one third the size to which you believe you should commit.

The next filter is equipment and premises. If the project you are looking at involves heavy engineering and you will work at it in your backyard, you have very little chance of getting by without adding a substantial amount of rental money for the right premises or yard. However, if you had those resources available to you in the core of your existing workshop, the matter can make your project eligibility better and reduce the amount of cash component in your offer.

Likewise, if the development process is going to require some specialist technology or equipment and you have already created this equipment in your company, then the possibility of ongoing capital is further reduced in risk. In this filter, some business owners will get taxation concessions against their funding of research and development. This should provide them a tax offset in excess of revenues the project could earn, given that most R&D projects and even projects in early-stage commercialisation does not have the profits to benefit from tax deductions. This varies across different jurisdictions, but might be something that you may wish to investigate.

The final filter, and the most important filter, is know-how. If you cannot approach any commercialisation project from a position of authority, then you just become an easy mark for money and there is a good possibility that the owners of the intellectual property would treat you as a bank rather than a partner.

Your collective knowledge and experience on the product subject, the industry, the market, and perhaps the manufacturing process, will further reduce your risk and add substantially more value to the money in the commercialisation process. This will not be recognised by the innovators at presentation, but will be appreciated quickly.

The final aspect of filtration is to have a checklist. You may have four key points that have to be met for any project to be invited to submit a single page proposal to you. On this subject, you should insist on a single page initial proposal for two reasons. Firstly, it is imperative to assess the ability of the innovators to present their product in a concise and descriptive way, and secondly, your time is valuable. In most cases, if you do not set a maximum proposal length as part of your filtration process you can hundreds of irrelevant pages.

The next part of a checklist is to look at your five elements for your project and make sure that you are satisfied that you can work with that. In dissecting the people (the first *and* last assessment element – for a reason) you need to look at background checks and history in commercialisation.

If they have patented any product involved in the process, you need to find where the funding came from. If the innovators owe money to third parties, they may have litigation pending in the wings and you need to be aware of this before it actually starts. In some cases, litigation may be pending and this will not show up on any credit report or court documents, but can be asked about in any questionnaire, before a secondary review of the product or process. If your questionnaire asks directly of "any known or potential litigation relating to the company, the project or the people involved" you may be given a surprising response.

Aggregation

The most fun part about becoming the magnet is your ability to sit above all the different opportunities, and calculate the synergy between sets of projects, the common aspects such as markets or industries, the synergies of technical personnel working within each of these projects and how you can merge two or more projects together, to form your own incubation laboratory.

It is fair to say that ego plays a big role in IP commercialisation and as the magnet, you have to assess the egos of all parties to establish if they have the potential to work together to create a cohesive team and project. If you believe you can get these teams together and accept that some key personnel may become redundant as a result of aggregation, then you are well underway to having a low-risk model for commercialisation with a ready market for trade sale or public listing.

The aggregation process is far more than just putting groups of people together and making them work. Firstly, there will always be an air of competition for resources and if this is not dealt with early, it will foster loyalties and competition within the group. The worst possible outcome would be that these groups start to compete for resources and may even sabotage each other in order to be the last man standing with a commercial project.

Aggregation suitability can include shared or common markets, similar skills or common plant and equipment requirements for the development. You should also look for potential integration of one to the other in the value process, and most importantly, economies of scale from development skills. You should not pass on projects as you wait for the perfect cluster. In my 28 years of commercialisation, I have not been able to get all of these to align in any single cluster of projects, but have created more rapid development times and bigger opportunities, through making two or more of these work in an incubation environment.

Start with the End in Mind

Author and business sage Stephen Covey once proposed that any project or negotiation should start with the end in mind. This is more than just setting objectives in the aggregation process for intellectual property. If you understand that your next project should be a trade sale to a particular public listed company, you need to understand what it is they will buy and what they might pay for it. This means you have to have an intimate understanding of their evaluation and selection criteria before you start your aggregation process, so you know just what your project could be worth when you meet your sales objectives and comply with their selection criteria.

Vague Higher Authority

Before recruitment of any intellectual property projects, we need to set up a vague higher authority. This is an imaginary panel of investors who provide the money and keep you accountable for performance of all the projects. If your innovators and project leaders are made aware that the buck does not stop with you and that you are accountable to an authority that they cannot approach, they are less likely to spend time and other resources arguing the point when they fall behind.

I have used this technique quite successfully in a lot of businesses (including as a landlord) where I talk about the investment syndicate and how callous some of the members are towards those who do not pay the rent on time.

Having an investment syndicate, which may or may not exist, is an excellent way of placing someone on the other side of yourself whom you can blame for an absolute lack of empathy in any project. Part of our problem with being human is that we have empathy and we find it difficult to distance ourselves from the tough decisions we sometimes have to make in business. If there is an investment syndicate that you supposedly meet with once a month, I suggest you schedule golf or fishing for the day after you do your Project Status Reviews, as the time you have to go to "meet the syndicate." You will find the project teams will work harder for you so you do not have to face a grilling because their projects didn't perform as they promised.

At this point, I should apologise to my intellectual property clients I have had over the last 28 years, who upon reading this, recognise that the syndicate that I have spoken of, didn't exist. In my defence, I saved hundreds of man-hours of argument by simply asking "how am I going to get the syndicate to buy that?" As an aside, I probably saved both of us a considerable amount of money and time, if there were a problem that I had to explain to a third party that the project leaders wouldn't be in front of. It was sometimes easier for them to fix the problem before I stepped out the door to my "syndicate meeting" with all of my PSR sheets under my arm.

IP Selection

The hardest part of all of this process is going to be the selection of the IP you want to incubate. In most cases you would think it can be difficult to find the perfect intellectual property that will complement your skills and your other projects you want to include in your stable. Nothing could be farther from the truth. Having a checklist in place before you start, will ensure that you do not become flooded with ideas, but rather your filtration will only give you the ideas that fit your selection criteria and will complement the other projects in your portfolio.

Once you set your filtration system, you then have to set your core theme that you are going to use for this incubation project. Most incubation projects have a life-cycle of about 18 months. Beyond this, you need to reject the project as what we term the walking wounded. This refers to companies who will probably never grow to the extent you need to grow, in that commercialisation period. Although not impossible, it is considerably harder to become a viable IP opportunity for other companies to purchase, if the project takes several years to incubate. Once you add the extended burn rate, these types of projects can send you under.

Just like staff you need to have your key performance indicators in place so that poor performance can be identified and rectified on an almost automatic basis. It is important that no project can slip below the radar or you may have dissent within the group where three teams are carrying one and like most entrepreneurs these people will not stay quiet for long.

Build Your Innovation Cell

When you decide to build an innovation cell, you need to have the structure and a set of rules in place so that ownership and control are very clear from the outset. Any intellectual property projects being accepted become jointly owned by you as the incubator and the innovator as the project vendor. It is always best to have an ascension model which recognizes contribution as unequal, and will allocate further equity in the project for meeting agreed milestones on a separate individual basis.

This means, that if the innovator achieves his outcomes before you both agreed to, he will receive his equity upon achievement of those milestones and you will have to work harder to balance your share of the project from the start. This works both ways, and ensures that nobody is going to rest on their reputations or the product promises, if they are to own any future part of the project at the value realisation point.

Your milestone key performance indicators form part of your incubation agreement with the innovator and should not form any part of a relationship with any other project in your incubation portfolio. It is important to silo each project, so that you can excise any underperforming project at any time, without affecting the relationship to others.

This becomes critical when setting budgets. Unless your innovation groups are highly disciplined and the innovators are prepared to learn at an exponential rate, if they do not apply the learning or the discipline you provide your project will start to become one of your worst nightmares: the cash burn. The assumption used here is that they have joined your incubation group because you have skills and resources they do not have.

Your cash burn rate is one of your most critical key performance indicators – for your IP teams as well as you - and in order to maintain this at an acceptable level, there may be times when individual project managers forgo their income for a week, because they've overspent on other areas of their budget. This will generally only happen once, unless the person is not a fast learner.

The carrot approach to this mindset is that if a project individually runs out of money, "your syndicate" will not fund them further without requesting that the innovator gives up additional equity to the syndicate. This applies a direct consequence to any poor administration or budget carelessness.

Project Status Review

The most critical area of project incubation is going to be the accountabilities. The most important accountability is always the project status review, commonly referred to as PSR. The idea of a project status review is for project leaders to share any issues that may have potential to derail the project, before these can. In order to make a project status review process fair, it would be prudent to include one member from each of your incubated projects onto the project status review committee. This ensures that the competitive bias is dissipated when all panel members know that they have their turn presenting their own PSR to the panel.

The project status review sheets should start at fortnightly, when there is a high activity cycle in the development, and moved to monthly when things stabilize and there is very little transition or change in the project. It's up to the project status review committee to determine how often the PSR reports should be handed in and all members of all teams should provide PSRs to their team leader before these are handed to the PSR committee.

If a project runs off the rails, the PSR committee would receive and interview the team leader on his PSR, but may also choose to interview all members of the incubation team, to identify where the problem lies and what can be done about it. It can be more of an advantage not to have the team leader sitting in on the individual PSR presentations by team members, as they are more likely to be frank about any situation that they would otherwise be hesitant to discuss in front of their managers.

The Walking Wounded

From time to time, up to 20% of your projects will not be meeting their milestones and may not be able to achieve the financial outcomes stated in the original proposal. When this becomes evident, it is prudent to isolate such projects and negotiate an exit for your incubation team, so that the budget for such projects can be reallocated elsewhere, perhaps even to a replacement project.

Before you exit the project, you need to interview each of the team members individually, to determine who is worth keeping and who is not. It could be that you can redeploy key members to one of the other projects that need the additional resources. One of the key benefits of incubation team management is that moving people around does have significant advantages in mixing up the thinking and keeping a fresh outside set of eyes on each project.

Separating Ownership

Although team members may work together and may share space and tools with other teams, it is important that ownership be kept entirely separate until the commercialisation is realized. This also suggests that ownership should not be cross pollinated, because if one project falls over it becomes an awkward situation when that team exits. This is worse when they have been granted equity in another project.

Keeping ownership separate also enables you to dispose of any particular intellectual property if circumstances make this prudent. It may become less attractive to your group or substantially more attractive to others, and you perceive that circumstances will change in the future. One of your greatest leverages is the power of options.

Having separate ownership and lowering the reporting and accountabilities, will enable you to collect and dispose of underperforming projects as may be recommended by your PSR committee, or even your "investment syndicate." Unless there is identified collusion of dishonesty, you should make judgments on projects on an individual basis and never collectively.

This means you also need to prove that you are not playing favourites and resources are available to all team members, so that decisions you make about disposing of particular teams, will be supported by other teams. There will be times when you need to dispose of projects which are flying high, but you are faced with an opportunity to realise a good return on your investment. Ultimately the return on investment is why you got into this business in the first place, so you need to be prudent about how you push on with this.

Periodic Value Testing

If you schedule your project status review meetings for monthly intervals you need to establish upfront to have a periodic value testing process. This will help to determine that yours and their investment is growing as originally agreed. Although you may have allocated a budget of say, one million dollars for the development of a particular intellectual property, the drawdown of these funds should be incremental and kept to a milestone achievement, which is judged in your PSR meetings. One of those milestones should be a periodic value testing, to ensure that you are not pushing a hundred thousand dollars a month, every month, into a project that will ultimately be worth less than one hundred thousand dollars.

Your value testing model should be agreed to at the beginning of the project, but needs to cater for differences in the stages of the development life cycle. This suggests that one way of testing value in a project at a very early stage, may not be relevant for testing a project at a later time or stage.

From Magnet to Magnate - Spinning Out IP

We have talked about exiting projects that are underperforming, but we also exit projects which outperform their agreed milestones, for positive reasons. As no two projects will progress along the same timeline, decisions must be made on a quarterly or half yearly basis on whether a project is mature enough to be spun out. One of the fundamentals in spinning out IP is never to think that you can do it yourself. You need to partner with somebody who has access to the markets, has an intimate knowledge of the industry, has the commercialisation and marketing teams available, can share the profits evenly and be accountable to you in your capacity as a shareholder.

Ideally, the commercial partner should be a public listed company, so that you can spin out the technology for equity in the company. This is usually done in stages. The first of these is the handover and the second is upon attainment of an agreed sales key performance indicator. If a project is worth $100 today there is a good chance that it will be worth $1,000 tomorrow, but you cannot prove this empirically.

The next best thing is to have the second tranche of value determined by a formula which is tied to the valuation multiplier that the target partner company uses to value the shares. This may mean that the first half of equity would represent $50 and the second half may equate to $500 of equity value, under this model. This mitigates the risk for the buyer and guarantees the recognition or realisation of the true value, by you the vendor.

Aggregation for an Exit

The holy grail of exits, for an IP incubation program, is the aggregation. This is where products or services that emerge from your incubation are so similar in shared markets and industries. This could provide sufficient turnover to deliver you a public listing or at the very least a reverse takeover into an existing public listed shell. This could be an innovative way of rewarding your team leaders with options and/or middle management roles in the public company.

In most countries, the process becomes an exchange of shares from one company to another and may not attract taxation until the shares are disposed of. This is something that should be taken up with your accountants before entering into any agreement for vending the IP. It is important to seek accounting advice relating to acquiring shares or options.

Maintaining momentum

Once you start to dispose of IP projects, you need to have a ready list of replacement projects in keeping with your team skill sets and resources that you have available. It could be that your half yearly periodic testing review includes a cursory look at some replacement candidates for projects you intend to exit.

Because there will always be an abundance of IP projects, it's up to you to pick the good ones and have them understand that the project would be in a holding pattern, until you are ready to accept them into your program. In most cases, the average innovator is looking for money and not help, but if he is not able to get money from others, then you stand a good chance of him being there as you become his fall-back position.

So having a waiting list and having the PSR committee vote on a waiting list could be very cheap and become an easy way of bringing new projects on where the risk is assessed by those who have to work with them.

What you do not want to get involved in is a Dutch auction of potential suitors and a game-show atmosphere of IP presentations, where you are told lies in the heat of the moment. In a competitive environment, IP owners may firmly believe that the next guy and the last guy both lied and they do not stand a chance if they do not lie too. Keep your opportunity funnel open and use your selection criteria as your first filter. It may pay to have an application form online, which requires those people to tick boxes as a qualification against your criteria, before they then give you, in limited words (without any attachments) a quick précis of their project.

Some members of your PSR committee may choose to do the first evaluation on some of these projects. They have their own vested interests in making sure that they can work with particular people, can add value and understand the technology in the new project.

I cannot emphasise enough that first touch-points with innovators should be through online forms and not in person. There are very few inventors that will take no for an answer and some see it as their life mission to convince you that you are making the wrong decision if you turn the project down.

If you are saying no to a person who does not know you, they would find it very hard to hound you into changing your mind. When it comes to inventions, sadly, "no" does not mean "no" to many inventors. I should emphasise that I do not see all inventors or innovators as this passionate or focussed. Some of my best clients have become friends and some of my best friends become clients, because they are fair, decent people with good logical understanding of what they have and what it might be worth to others.

Fairness

If you have elected to establish an IP portfolio management process and are running an incubation program, the onus is on you to become fair and equitable in all your dealings with the principals of each of your projects. Some portfolio managers, in particular of the venture capital groups and private equity funds, tend to get greedy with these projects and start to charge fees for administration, accounting, PSRs and other interactions.

Some investors do this because the people they are dealing with are more vulnerable than others. Some may have spent their life savings to get into the project to the point where they presented it to these incubation groups and were accepted into the program.

Less-scrupulous incubation managers seek an opportunity to take advantage of some of these innovators, but these investors do not end up with a lifetime of innovations, bouncing across many industries and markets, with just one day a month involvement in the projects that make them the most money.

Remember this key point when you calculate what the project is worth to you, which is usually at disposal.

CHAPTER 12

Value from Expertise

"If you teach somebody what they need to know,
you earn the right to help them implement it for a fee..."

—Frank Kern

This chapter is specifically for IP owners who have developed a service or a particular expertise and believe others would pay to learn or use this. Sometimes this service or expertise is based around knowledge, experience and know-how, but may not be able to be patented or protected.

For anybody who had driven their business to be an industry leader, there is a strong possibility that they have developed some expertise in the form of knowledge that can reduce the development or learning interval by others seeking to enter into the industry, by as much as 90% of this learning interval. If this equates to a reduction in developing expertise from ten years, to one year, what is the value of all of those years of income? You may have many clients that pay for you to come in and set their business up to run like yours. It may be that there could be hundreds or even thousands of consulting opportunities like this, if only you could be everywhere at once.

The three best ways to achieve this are to (1) franchise the model, (2) to hire and train others to deliver your programs (deputise) or (3) to monetise the learning and deliver it on a broadcast medium like the Internet. The licensing model has been addressed in detail in an earlier chapter, but the key drawback for unprotect able expertise is that you cannot hold these deputies for very long. Once you have trained them, most would drift off to start competitive practices and soon you have competitors whom you have trained in this area.

We are all familiar with franchising. This is ultimately how we build a model around our collective experience and share this with others. We have a structure and processes for them within which to operate and people are prepared to pay hundreds of thousands of dollars to step into this "made-for-you" model of business. Many will pay a substantial premium for this, as they have significant risk-mitigation through delivered expertise, long-term coaching and shared promotion. For the franchisor, there may be protection advantages under the franchise contract, but some of these might not be enforceable beyond the franchise contract period.

But is this the best use of your expertise? Imagine the thousands of franchise management personnel that are needed to hold all of that together? This is money generated from your innovation, which goes to wages for people to keep it going. Not very much left for you....

The best way to capture and commercialise expertise (including, knowledge, experience and know-how) is to build a monetisation model around it. If you can share this with someone and have them make a substantial increase in their income, then the only thing that stands between you and a fortune is a one-to-many service delivery process. The only way I know (from 30 years of commercialisation experience) is to tangibalise what you have as a training package, that can be delivered automatically, one-to-many or as a hybrid of both options.

My Story

Nearly 30 years ago, as an undergraduate with a background in services marketing, I was approached to manage the intellectual property of the University. The intention was for me to generate business opportunities for the intellectual property and have the technical people close the deals. Unfortunately, most of the technical people had no understanding of business negotiations and eventually I ended up in this position by default.

By 1987, I was commercialisation manager of the intellectual property emulating from this university. Eventually, with the 1987 stock market correction, a lot of industries stopped investing in intellectual property for a while and the department was wound down. I was offered a 12-month contract to finalise the last eleven projects and this was the start of my consulting business.

The business itself provided a reasonably good living, but every time I tried to scale up, I recognised that I had specialist expertise and had to train someone extensively to at least be able to sell the services to others. This became a cycle. I would recruit and train competent people, but within six months they considered they knew enough to own their own business. Most of these would bounce out after twelve months into their own practice, taking one or more of my current clients with them. The net benefit over a 12-month training cycle became negative for me. This was without factoring in the additional competitors made for myself over a ten-year period, before I gave this up.

My decision not to having successors, meant that I knew my business would eventually die with my retirement. It was on track to provide a reasonably good living for the interim, but I had no way of monetising at that point. The Internet changed all this. The model I needed became a combination of what I built and what I co-designed with a business partner.

Several years ago, this colleague of mine who worked with me some twenty years ago on another project, arrived back from Sydney to work in Western Australia. He didn't have a business and didn't quite know what to do and so he approached me and asked me to find him a business, train him for a business, or build a business with him. We eventually decided we would plan this business out and started a joint business together. This gentleman was a fast learner highly intelligent and was not in any way swayed by the business protocols of the day. He would prefer starvation and sleeping under a bridge than become an employee and he made that very clear.

After several months of lunches (we met once a week at a cafe, to discuss what sort of business would be ideal for him) we built a management consulting practice around an ascension model. This started with some free and affordable ideas for implementation, but would see businesses recognise the value we can offer them, before they contract to us for long-term change within the business. Our premise was to *"....double your net earnings in ten months - or your money back."* At the time, there were many promises being made by others, of ten times your business in ten weeks and double your business in six weeks, but we knew that only incremental change would be sustainable and we wanted these businesses to be so successful that they would bring us our next customers.

That business is at www.formula1forbusiness.com and is an online management consultancy practice for service based business owners in the one million to ten million dollar earnings bracket. The ascension model we use will be explained in greater detail, to give a better understanding of just how ascension modelling works.

The concept of this ascension model with a grow-out service based business, was easily adaptable to my own practice and I am very grateful for the time Simon and I spent developing this model and refining it to a point where we knew it would fire every time. This has enabled me to have a practice which I can share with others and more importantly, sell, when the time is right. This is achieved because I use the collective expertise, knowledge and know-how, to monetise the learning and share it with one-to-many, without restriction on the geographic territory.

The Monetized Membership Model

There can be many configurations of a monetisation model. As a learning exercise, you need to build a basic framework which can be adapted as you build the model. I can suggest that the fewer bells and whistles you have on your monetisation model, the easier it is for the buyers to understand and the easier it is for the customers to buy. The four core layers of a monetisation model in service industries are (1) where clients or prospects can read about it and implement what they learn, (2) where you present your ideas to groups in seminars and webinars, with an audience of clients who are on a similar learning curve, (3) where you offer clients and prospects help with some hourly-rate teaching, coaching and guided implementation, and finally, (4) made4you implementation, where you build them what they want and teach them how to drive it, usually for a fixed fee. This last point may sound a lot like franchising, but it generally does not provide a trailing income after you deliver the service.

The two key issues for all the options in this business model are based around pricing and time. The less time you spend with clients, the more clients you can have. However, the more you charge clients the less demand you will have. It takes some careful economic modelling to identify where your sweet spot is, but in most cases, the best is a one-to-many broadcast learning with some limited one-on-one by phone or Skype. This eliminates travel time and gives you the option of face-to-face meeting with only those that need it. The standard practice model is to have a "one-to-many" session per week of Q&A and at least one live learning session per week. You must also have a library of relevant tools and learning that others can search through, to bring them up to where you are on the development journey. Finally, you need to have troubleshooting tools to identify if people are falling behind or not implementing, because these are clients that you will lose over time.

Giving "one-to-many" learning means you can reduce your price substantially, making your process affordable to thousands and not restricting you to a few dozen clients. As you develop this practice model, you can deputise some of your senior members and have them run one or more of the weekly Q&A sessions, particularly if their expertise is in the area that you are promoting for that week or month. This allows expansion geographically and demographically, as you apply your tools to different industries.

The single most significant competitive advantage on this model is affordability. It's easy to show prospects how they can reduce their learning from ten years to one year with what that will equate to in years of extra earnings. When you add on the additional business value you generate from having a more substantial business trading over several years, their sale value of their business could be more than double. In most cases, people want "one on one" service but are reluctant when they are faced with the real price of this service from a true expert.

If you were to offer someone assistance for $500 an hour (or a pre-paid block of ten hours for $4,000) compared to shared learning at $1000 a month, it will not take them long to realise where the value really lies. Eventually, you will continue to increase your hourly rate until everybody refuses the hourly rate and tries the monthly rate. Ultimately, they prove to themselves that they can learn this way, effectively freeing you from visiting clients and spending one-on-one solving the same problems week in and week out. You might recall we refer to this as the FOP pricing model (F… Off Pricing).

If these people are faced with a pre-purchase of a block of ten hours for say, $4,000, would they opt for the $1,000 per month on a 12-month contract, if they are offered a money-back guarantee within the first month of service? This hypothesis is very easy to validate.

Every model has bottlenecks. For an on-line service delivery membership model, it may be all about the speed of your Internet access or the number of touch-points you build in, that will restrict how many people you can offer these services. Working the absolute one-to-many business model, you can host Q&A sessions for thousands of members, then post up the recordings of these sessions for those of a different time-zone, who may have found it difficult to participate live. If you have awesome bandwidth in your region and you can hold a one-to-many webinar with hundreds of participants (in listen-only mode) then you have taken the first steps to eliminate the biggest bottleneck in any service business – the principal.

One of the fundamental requirements of a one-to-many model is the necessity to strip out all the local content, so that it does not preclude members from other areas. In my own model, I had to remove all references (other than generic reference) to business and research grants, as these differ across jurisdictions. You may have a region-by-region library of these available on-line to download, but keeping this up-to-date will be difficult, unless your members provide the material, over time.

This provides an outline of the membership model, where you earn the right to charge members a long-term membership, usually payable monthly and renewable annually, to access your learning and advice. You work your way through the business model in a relevant sequence and they will take the journey with you, applying the weekly learning to their businesses.

One of the keys to a proper membership model is the recruitment of members. This falls into two core components. The first is Lead Generation, where you find a "hungry herd" of prospective members who can identify with the problems you solve, and the second is Lead Conversion, or converting people who have shown interest in what you offer, into paying members. We will look at lead conversion first, as this is best explained as the Ascension model of on-line client recruitment.

Concurrently, service providers should still search for new prospects from existing members and through all types of channels, including sharing the stage in collaborative events to address other service providers' hungry herds (for a percentage of your fees) as a supplement to the Ascension model.

The Ascension Model

The objective of a basic Ascension model is to convert qualified prospective clients from observers to paying clients, in approximately 4 stages. The Formula1 For Business model uses five stages in its conversion, but is earning money from the last two stages. It is important to maintain these stages in sequence.

The first objective is to have the people who are searching your site, finding and reading the information on other sites which you used to promote your business or run your banner ads in. When you have these people register on these different sites, you can provide an immediate download of a short piece of information that they will readily recognise could deliver them an immediate saving or benefit. This is sometimes called the lead magnet. We usually use a number of different techniques including small ads on Facebook or LinkedIn, which lead to another website, known as a landing page. This may highlight the need in a very short video, usually less than two minutes, and offer the visitor an opportunity to download a short piece, such as a check list. If they download within minutes, they can implement within hours.

For an example of such a landing page, I refer you to one of my landing pages which provide such a checklist www.makemyinnovationhappen.com. The video on this page is a whiteboard scribble, but the key to it is the voice-over. For people landing there for the first time, they hear the voice of somebody famous, telling them what could happen if they do not have this checklist. The novelty factor alone enables this to be almost viral, as despite the intensity of the message, people are amused by the delivery and send it to their friends.

Once the viewer connects with the problem and recognises that the download will have a solution, the only thing preventing them downloading this for free, is the hesitation with you asking them to input their contact details. This is the first stage of trust I ask of people when they know nothing about me or my services, but must make a judgment as to whether or not I'm safe to do business with. They have to decide to give their email contact or phone numbers to a stranger. To lessen the risk for these people, I make it clear that they will not be asked for money.

The first transaction is an exchange of contact details, for a piece of information that will make or save them money. From here, once this information proves its value, they will want further information on the systems and processes that I offer. They trusted me with their email address, and they received some information which made them money. I have now built a small amount of trust with those prospective clients.

Within hours, they will receive an email from me which thanks them for their trust and offers them an opportunity to follow-up, with a free e-book, which is particularly relevant to the stage immediately after the checklist they have received. Again, anybody who has received an immediate benefit from the checklist they downloaded, will be hungry for a free course designed to help them transition through to the next phase of the commercialisation journey.

That next phase is the offer of an on-line course, on the commercialisation process of intellectual property. If they sign up for this course, they are told at the beginning that this is the first module of a four-module course and they will be offered the other three parts for a fee. They are assured that they do not need to do this and that they should get the benefit immediately from the first week that they do. If they choose to walk away at the end of that week, they are free to do so and there is no contract between them and myself at that point, for any services. This is the second stage of trust and the second opportunity for them to receive a substantial benefit with no exchange of cash.

These automated online course modules are provided to the prospect on a daily or weekly basis, with a handy note of encouragement and an opportunity to solicit some feedback on their progress. They are always reminded of the content for the course beyond what they're doing, so that there are no surprises when they have completed their free on-line course. Each message is seeding the full course and emphasising the value they will be able to get from this, in comparison to the value they getting from the free course right now.

At the end of the free online course modules, they receive an invitation to sign up for the next three modules of the course. They are given details of each module, the outcomes these modules should deliver, as well as the direct benefit to them as the business owner or intellectual property owner. They are offered a significant discount if they accept this course within 24 hours of completing the other online free course. They are offered easy payment terms and provided a table of proposed contents, over a three email sequence, for the 24 hours after completing a course.

For those that sign up for the course, they are offered membership after this, with a discount on the first month, commensurate with the course fee. This is a 12-month engagement and by that time, they understand what they are going to receive and the benefit that is going to give them. To date, they have received substantial value in both the free and paid-for publicity, and have expectations that the membership will deliver more.

For those that do not sign up for the course, they are engaged on a series of six emails over an eight week period, which encourages them to become members directly, and start to benefit from the information they learn, in the time that they will learn it. All the emails are automated and managed by an overseas-based assistant.

The ascension model is based around the statistics of acceptance. For those that do not progress from early times through to later times, there are newsletters, monthly updates, encouraging emails and short question emails, to stimulate some interaction and to keep their interest in the material. Some of these are later invited to webinars and seminars, in order to experience the service more and become members.

The alternative to courses is the webinar and seminar model, where people accept the checklist, are invited to attend an online seminar or a larger seminar in their town or city, to share the experience of the service provider in a public environment. In years gone by, this was a very effective tool to recruit members and there are still many seminar speakers who create an urgency to discount and limitation, to have members of the audience rush to the back of the room with their credit cards, to sign up for a 12-month program or similar.

This "rush to the back of the room - only 50 places available" model has been a little overdone these days, and has always been particularly hard if there is a number of different speakers, as people who are speaking last are talking to an audience who have already committed their credit cards to one or more of the previous speakers, perhaps on a long-term course or membership.

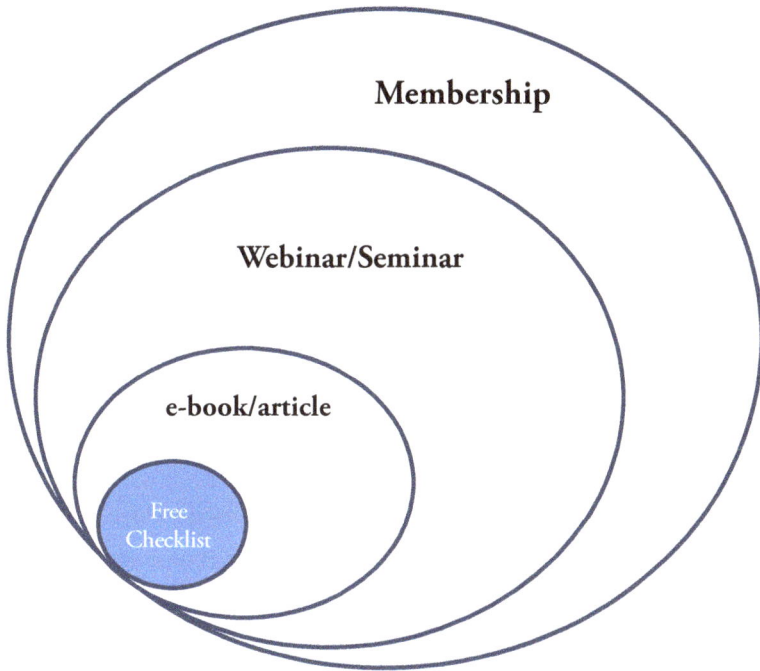

Lead Generation

The process of lead generation for online services is about finding and corralling a hungry herd of prospects that want and need your services and can afford to pay for them. Your first objective is to have these people give you their contact details, so they appear in your database and receive quality information from you, which they will see as immediate and substantial value to them.

There are hundreds of ways of generating contacts, both online and off-line, which will ultimately provide you with a list of prospective buyers of your service. This is sometimes referred to as the hungry herd. The first objective is to own the hungry herd, by having these people enrol into your database, by requesting some free information that could benefit them. They have effectively enrolled in a process to receive information from you in the future, as well as receive something which has delivered substantial value to them, thereby building trust and credibility for you.

As you build your hungry herd, you should be mindful that you do not have to initially sell them anything, to generate sales revenue from the followers. If you become the online leader in your industry, people will take a recommendation and in most cases, accept invitations you provide to them through third-party collaborations.

If for instance you have 100,000 people who have joined your list to receive information on your topic, you may sign a collaboration deal with a third party who delivers a course or membership, whereby you will collect 50% of the fees from all signups of members from your site. With a .05% conversion on an annual membership of $5,000, a 50% collaboration fee could net you $1,250,000.

Just being the leader of your space can generate substantial revenue without any product or service monetisation. Of course, one must be careful not to overheat the list. If your hungry herd is fed too many offers, they grow tired of your recommendations and will opt-in for less and less over time. For more information on how to build the hungry herd and become the leader in your industry, you should search out the book "The Digital Delusion" (Buehler, 2014).

APPENDIX A

A Pre-Venture Capital Checklist

So by now you've got the right business structure set up, the products are starting to sell, you are experiencing rapid growth and you realise you are now in a position where you need to attract major funding. Without funding you might not be able to take advantage of the demand you have identified in markets you cannot afford to reach. It is time to plan your venture capital presentations.

The following is a checklist of what you will need to have completed or achieved, before you plan your venture capital presentations. It is very important to understand that you only get one chance at any VC presentation because they will see hundreds of presentations a month and will only allow perhaps 5 to10 groups to present to them. If you present too early, you get a cross instead of a tick and you will not be invited back.

Remember, most of these people are human too, and may not directly say go away. They might give you some words of encouragement but essentially they have given you 30 minutes to present and you have failed to set them on fire, so they are looking for the next project. The big lesson in preparing for a venture capital presentation is to be absolutely sure that you have everything they need, so you do not blow your one and only opportunity with them.

The following is a list of the different aspects of venture capital. This includes what those guys will want to see in the information memoranda as well is the PowerPoint presentation. This is designed for you to meet what they need for a rapid analysis of what you have and what you are offering.

Your Pitch Deck

Your pitch deck is a PowerPoint presentation of your proposal to venture capitalists or other investing parties, in the third round of capital investment. The first point I need to make on presenting to venture capital analysts or potential is that most of these people see a lot of PowerPoint presentations and if you present everything, they are going to get bored in the first five minutes and look for ways to reject you so they do not have to put up with your presentations in the future.

The smart presentation uses fewer words and more pictures. Each frame should be a talking point rather than a summary of what is being spoken. Certainly, one should never put a slide up and read out the contents, as this is tantamount to saying that you do not believe these people can read and so you are going to read it out for them.

There are usually just ten to twelve slides required and if you start to press more slides into the pitch deck, you run the risk of boring these people or becoming repetitive. They do not want to have the details presented to them. They are looking for a précis of highlights of your project and business model process, for them to decide you are somebody they can work with in the future.

Remember, the first and last decision point for them is the people. They need to know that they can be comfortable with you and if you try to commit death by PowerPoint, their tolerance level will reduce substantially in minutes. It does not matter how good your product is, if you take twenty minutes of talking about yourself and your team, before you get to what the project is you will have already lost them.

There are different schools of thought on PowerPoint presentations, as the most effective for people who see more than one presentation a day for all of their working lives. These need to have no more than three words per line and five lines per page. The most effective presentation will always be 3 to 4 words per page with a photograph. Treat your PowerPoint slide as a talking point, not as a reading guide.

All the technical data should be in the handouts and if you were to breeze through this PowerPoint without any questions, it should take no more than ten minutes. This suggests one minute per slide but in reality, some of the slides, such as product, would generally take 2 to 3 minutes whereas team will take 15 to twenty seconds as you acknowledge the people from your team in the room and mention the others. Here is a sequence of the appropriate pitch deck for a venture capital presentation:

1. Project Name and Team
2. Summary
3. What we have
4. Who wants it
5. How it works (1 slide)
6. How we own it (uniqueness)
7. What we want from you
8. Why YOU should buy into this
9. What you will get
10. Expected cash flow (with break-even) – just quarterly summary
11. When you will reach the exit option
12. Where to from here

Cash

There are essential boxes which need to be ticked around the explanation of the cash slide. These would include your cash position and the projected cash flow for the project for the next twelve months and beyond.

Present this as a one-page table in PowerPoint, as a short summary, of quarterly columns and a number of rows such as sales, cost of goods sold, gross revenue, expenses, salaries, capital expenditure and net revenue. This can be a tidy little one page matrix, which shows them at a glance what your cash position could be, given a solid source of funding.

You need to have a comprehensive cash flow statement as well is a profit and loss statement, including at least twelve months historical and three years projections. This should be in the information memorandum that you leave for them after you present. As mentioned earlier, I tend to deliver that several hours later, so I can include the PowerPoint presentation with a set of questions and answers, which were raised in the presentation itself.

Venture capital analysts will be looking for any unmanaged debts you have and will more than likely asked the question. Most early-stage commercialisation projects have considerable debt, as this is directly associated with a high-growth program. What the VC will be looking for is the source of that debt. If the principles have loans to the company and expect to get that money back in debt repayment instead of equity, it presents a big red flag to their commitment to the project.

They will also be looking at how you've factored in the payment for protection. Patents are not cheap. You may require patents or trademarks and you are continuing through the process with several years to go before you have full patents issued in the markets you've targeted. Prospective investors will ask how you intend to meet these payments and what would be the consequences if you were not able to meet these.

Be careful how you answer this. There are some analysts who may suggest to their boards, that they wait until this patent has lapsed before they do business with you. This gives them a competitive advantage in the negotiations as the technology may not be owned. More than likely, the tactic used by the less scrupulous business analyst is to delay negotiations until the last month of your provisional patents. You then have time pressure for decision-making which they can leverage against.

The Company

1. What is your company's mission? (How will you change the world?)
2. What is your legal structure?
3. What are the products or services your company delivers? What new ones are planned?
4. What problem or pain does your products or services solve or resolve (if any)?
5. Who experiences this pain or problem?
6. How has the company funded its development so far?

The Market

1. How does your company make money?
2. What is the value chain in the market? (Distributors / Suppliers / other key decision makers?) How large is the market? How could it be made larger?
3. Who are the target customers?
4. How do you reach out and get customers? How will you do this repeatedly? What prices do you charge? What price do your competitors charge?
5. What are your competitive advantages? (Why are you better than your competition?)
6. What are the barriers to stop new competitors entering your market?
7. How do you meet your ideal customers' needs better than your competitors?

The Team

1. Who owns the company?
2. Who owns what? Who is owed what?
3. How is your company shared between the various shareholders now and in the future? Who are the key team members and what functions do they perform?
4. Are there any members of the team critical to producing the products or services?
5. Who manages the Sales / Marketing / Operations / Production / Administration / Finance? What experience do your team members have?
6. What gaps do you have in the team and what steps are you taking to fill them?
7. How critical are these roles now and over the next 1, 2 and 3 years?

The Operations

1. How are your products or services delivered?
2. How long does it take to create and deliver your products or services? How much do they cost to deliver?
3. What intellectual property is owned by your company?
4. How do you protect your intellectual property?
5. Do you have competitive advantages within your operations?

The Finances

1. How much will the company make in the next 1, 2 and 3 years conservatively? How much will the company spend in the next 1, 2 and 3 years conservatively?
2. Of the money spent, how much will be invested in setup costs, developing intellectual property and other forms of investment versus just keeping the business running?
3. How much money are you seeking this round? What are you offering in exchange for this investment?
4. Types and quantity of shares
5. Percentage of equity pre & post-investment
6. How much more will be needed to complete the project for everyone?

Their Internal Questions

1. Who do we know in this space? Can we bring customers or suppliers to this deal?
2. Who are the competitors? How big are they, relative to this player?
3. What will their reaction be to this player in their space?
4. What other products or services will buyers use if this is not available?
5. Do buyers have to buy? Do customers have to use?

6. How much is the average marketing budget in the industry at the moment? What percentage of sales would this be?

7. What is our exit? If this does not happen, what is our contingency?

8. How cooperative are the current team?

9. How competent are they expected to be in the big league?

10. How suitable are they for Board Members?

11. If we do not buy now, will someone else buy? Can we buy later for cheaper? Can we still retain the same value?

12. What quality and level of reporting are we getting in the due diligence period? Can we expect these guys to report to us regularly?

Summary

The biggest question of all is the one they will ask themselves for every proposal. That question is "what if what you claim, or what you want, does not happen?" This is more than just looking at what you have planned as a contingency. They want to see what the consequences would be if things did not work out, just as much as to test how much you looked at this as an option.

They will be looking at the downside of not having sufficient capital, or if there is an over-spend or other issue, such as industry regulation or certification fails, or is delayed by several months or years. Given that money has a time value, this would eat straight into the bottom line and erode the profits they would predict for this project, which are already far less than what you predict.

They are also testing your frankness and reality and failures. It is great to be optimistic with your project, but for an analyst who has to commit millions of dollars to a project to make it work, there are inherent risks that will reflect upon him or her, if the project does not live up to their projections.

Everything you present is only used as a guideline by an analyst and in some cases their cash flow projections are much higher than yours, because they know they have connections which would give them early business. In most cases just like in the tender process, the person doing the evaluation has more to lose in career or trust, than anybody else in the transaction process. With this in mind, you should have carefully structured contingencies, which you answer upon questioning but not necessarily include in the projections.

I once did a presentation to a U.S. venture capital firm for an educational toy patent, developed in Australia. We did our best-case and worst-case scenario presentations parallel to our expected cash flow projections. We knew where the markets were and how we would get those markets and the most volatile factor of projection was the foreign exchange rate. At the time, the Australian dollar was 62 US cents and we knew that manufacturing in Australia for export to the US was going to be a very profitable exercise. Our worst-case scenario was factored in at $.80, as a measure of conservatism.

The venture capital firm had done their calculation at AUD$1 to USD$1 and after less than five years, this proved to be the case. Luckily, that toy venture recognised the volatility of the Australian foreign exchange rate and moved the plastic injection moulding operation to China. This was already factored in as part of their contingency, but the loss there was the claim of "Made in Australia", which was held in high regard by American mums.

-oOo-

APPENDIX B

Network Pre-Meeting Checklist Form

Name:	**Daniel O'Connor**
Qualifications:	B.Bus, MBA, CPM, FAICD(Dip), MAIM, AIMM, MAIeX, AIMCM
History:	30 years IP commercialisation - 28 years private practice
	Lecturer, author, facilitator, coach, consultant, project leader
	Company Director – private, public, listed and unlisted, Australia and Asia
	Member: UN Taskforce on Innovation and Competitiveness (ESCAP)
Looking for:	Inventors with a patent or prototype (product or service) who may have lost traction
	IP development team leaders
	Clients of patent attorneys
	Clients of grant writers
My services:	I get traction in IP commercialisation by cutting years to months
	I show people the 6 roadblocks and how to overcome them
	I show people the 2 things more important than the product itself

I structure projects to attract capital and help inventors to raise it

I plan and drive the project through to their value realisation point (reverse take-over, stock exchange listing, trade-sale, etc.)

I have a 6-week course on *Accellerating IP Projects* ($995).

I have an affordable monthly members-only subscription website for sharing my tools and providing regular direct assistance to clients on their respective projects.

My Websites: www.commercializeIP.com
www.makemyinnovationhappen.com
www.billion-dollar-napkin.com
www.takemyideaglobal.com
www.profitfrompatents.com
www.incub8IP.com

My Contacts: 0417 956 433 (Australia)
danielinperth@gmail.com